Electronic Circuits

Volume 1.2

Disclaimer

Electronic Circuits

Volume 1.2

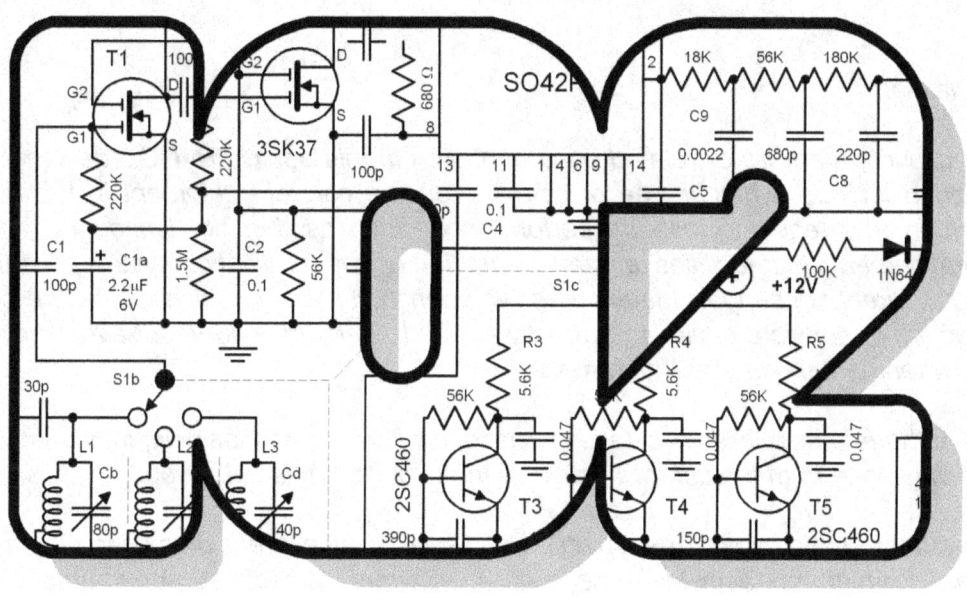

Intellin
Organization

www.intellin.org

Published by the

 Intellin
Organization

www.intellin.org

Disclaimer:

The circuits, software or related documentation in this book are NOT designed nor intended for use (whether free or sold) as on-line control equipment in hazardous environments requiring fail-safe performance, such as, but not limited to, in the operation of nuclear facilities, aircraft navigation or communication systems, air traffic control, direct life support machines or weapons systems in which the failure of the hardware or software could lead directly to death, personal injury, or severe physical or environmental damage ("high risk activities")

The author(s) and publisher(s) take no responsibility for damages or injuries of any kind that may arise from the use or misuse of the circuits in this collection.

The author(s) and publisher(s) specifically disclaim any express or implied warranty or fitness for high risk activities. The circuits, software and related documentation are without warranty of any kind. The author(s) and publisher(s) expressly disclaim all other warranties, express or implied, including, but not limited to, the implied warranties of merchantability and fitness for a particular purpose. Under no circumstances shall the author(s) and publisher(s) be liable for any incidental, special or consequential damages that result from the use or inability to use the circuits and software or related documentation, even if he has been advised of the possibility of such damages.

ISBN 1-4196-4622-2

Congratulations for having the third volume of
ready-to-apply circuits. You got the luxury of being able
to design and assemble electronic modules fast and
worry free. It is a sure way to optimize satisfaction in
your hobby. If you are a professional electronic
designer, it will help you beat the competition. Speed,
efficiency, short development periods, error-free, user
and maintenance friendly: these are the factors critical
for success. This invaluable book filled with 102 practical
ideas will help you beat project deadlines. Make your
ideas work!

Make your creativity pay! All that JUST IN TIME!

        informative...
                practical...
                        professional...
                                versatile...

PREFACE

Acknowledgments

Many Thanks to...

Engineer Mischa (Optical Recognition)
Engineer Salinger (Electronics)
Engineer P. Schmidt (Cybernetics)
Engineer N. Lay (Robotics)

INTRODUCTION

This collection contains 102 practical circuits grouped in nine general applications. Since most of the circuits are not limited to a single application, a circuit may have found its way into another group. This is one proof of the versatility of the circuits. Creativity needs versatility . You can combine several circuits into one large module to create a powerful electronic device specially designed for your exclusive project.

The table of contents lists the groups and titles of the circuits. The page number where a group begins can be found in this table. To find a particular circuit, turn to the group's beginning page. On this page, the circuits are again listed with their page numbers. To quickly find an application group, use the black markers on the edge of the pages. These markers coincide with the markers in the table of contents.

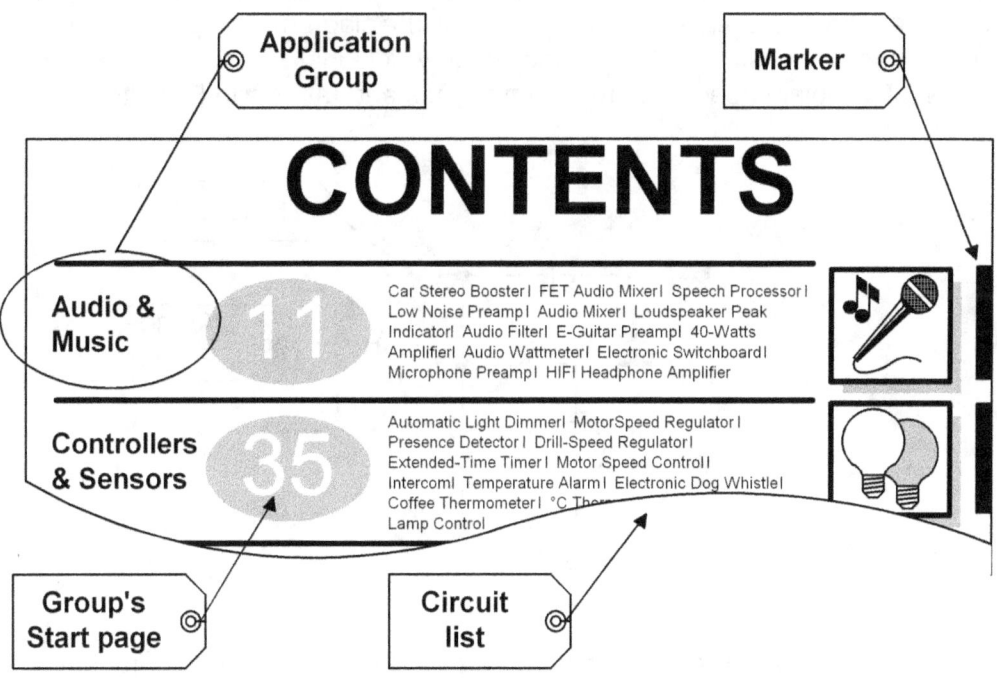

Electronic Circuits 1.2

The transistors used in the circuits have more than one possible replacements.
The pin designations are also shown in details. This feature can help avoid unnecessary
delays. The pins shown are either in the bottom view or front view of the transistor unless
otherwise noted. Large transistors which cannot or not planned to be installed directly
on the PCB must be installed on a heatsink. A dashed circle around a transistor means
that the transistor must be heatsinked.

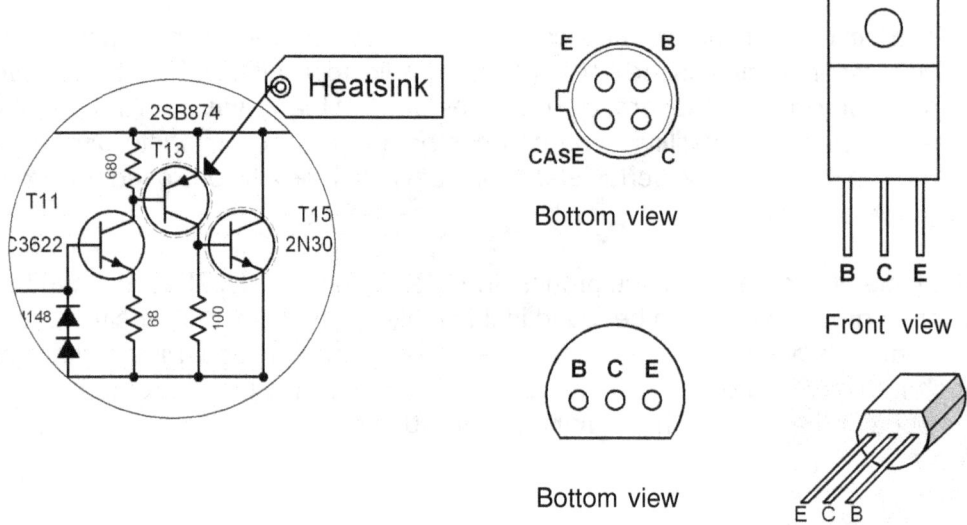

The capacitor values are given in microfarad unless otherwise specified. Electrolytic or
polarized capacitors are marked with a plus sign in the diagram. This plus sign coincides
with the capacitor's positive polarity in the circuit. Additionally, their voltage ratings are
also given. Nonpolar capacitors are ceramic types and rated with 50 volts.

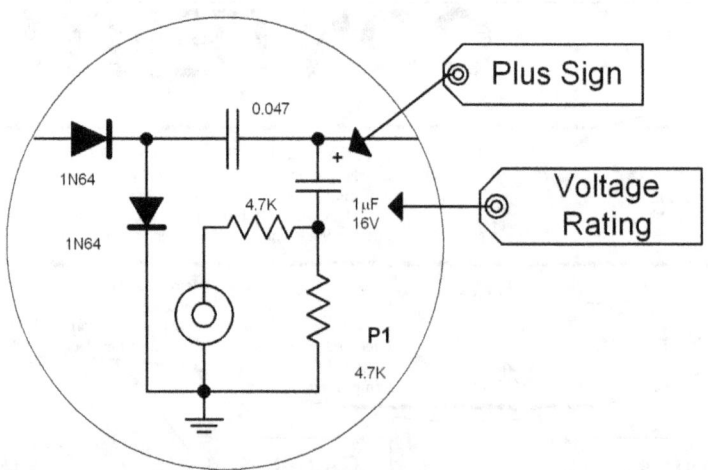

The resistor values are given in ohms (Ω), rated 1/4 watts and are of carbon film type
unless otherwise specified.

CONTENTS

Audio & Music  **11**
Car Stereo Booster-FET Audio Mixe-Speech Processor Low-Noise Preamp-Audio Mixer-Loudspeaker Peak Indicator-Audio Filter-E-Guitar Preamp-40-Watts Amplifie-Audio Wattmeter-Electronic Switchboard-Microphone Preamp-HIFI Headphone Amplifier

Controllers & Sensors  **35**
Automatic Light Dimmer-Motor Speed Regulator-Presence Detector-Drill-Speed Regulator-Extended-Time Timer-Motor Speed Control-Intercom-Temperature Alarm-Electronic Dog Whistle-Coffee Thermometer-Telephone Lamp Control-°C Thermo-Sensor

Radio Frequency  **51**
Cable TV Amplifier- 2-Transistor Radio-Sensitive FM TunerICW Filter-Pin Diode Switch-Field Strength Meter-Mini AM Radio-Ocean Emergency Radio-FM Antenna Amplifier-Morse Code Filter-Shortwave Preselector-UHF Antenna Amplifier

Hobby & Shop  **67**
Nicad Battery Charger-Thyristor Signal Generator-Frequency Divider-Regulated Mic Preamp-Superzener Circuit-Audio Delay Line-Trigger Switch-Frequency Multiplier-Electronic Potentiometer-Electronic Selector Switch

Power Supplies & Chargers  **81**
Variable Power Supply- 3-Watt Amp Power Supply-0.1V-50V Power Supply-Battery Monito-Nicad Battery Charger- 0-50V/0-2A Power Supply-DC Motor Speed Regulator- Voltage Monitor- Electronic Fuse-Current Alarm- 78XX Regulator Monitor - Voltage Converter- TTL Power Supply Monitor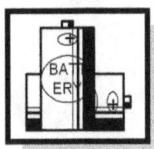

Digital & Computers  **101**
One-Chip Baud Generator- RS-232 Switch-A/D Joystick Converter-Serial Data Converter-Keyboard Doubler-Printer Computer Switch- Serial A/D Converter-I/O Scanning Keyboard- Parallel-Serial Converter- Video Signal Mixer-CPU Clock Generator- 8 Bit A/D Converter

Oscillators & Generators  **117**
Squarewave Generator-Digital Sinewave- CMOS Pulse Generator-Wien Bridge Oscillator-Digital Squarewave-Sawtooth Generator-48 MHz CMOS Oscillator-Power Multivibrator-Sawtooth Converter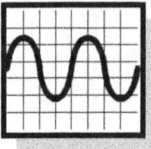

Testers & Multimeters  **127**
Acoustic Logic Probe- FET Voltmeter- Lamp & Fuse Tester-Continuity Tester-Audio Signal Injector-Analog Frequency Meter- Transistor Tester-FM Transmitter-Servo Tester-Digital Sample & Hold-Infrared Detector-Mega-W Multimeter- Car Thermostat- OPAMP IC Tester- Mini Frequency Meter- Trigger Amplifier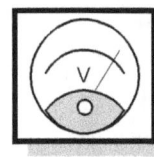

Auxiliary  **143**
Digital Bike Tacho- Auto Car Warning-Diode Thermometer- Diode Thermometer-Battery Monitor- Car Antenna Amplifier

This page is intentionally blank.

AUDIO & MUSIC

12 Car Stereo Booster

15 Fet Audio Mixer

17 Speech Processor

18 Low Noise Preamp

19 Audio Mixer(emitter driven)

21 Loudspeaker Peak Indicator

23 Audio Filter

24 E-Guitar Preamp

26 40-Watts Amplifier

28 Audio Wattmeter

29 Electronic Switchboard

32 Microphone Preamp

33 HIFI Headphone Amplifier

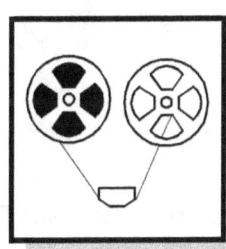

1 CAR STEREO BOOSTER

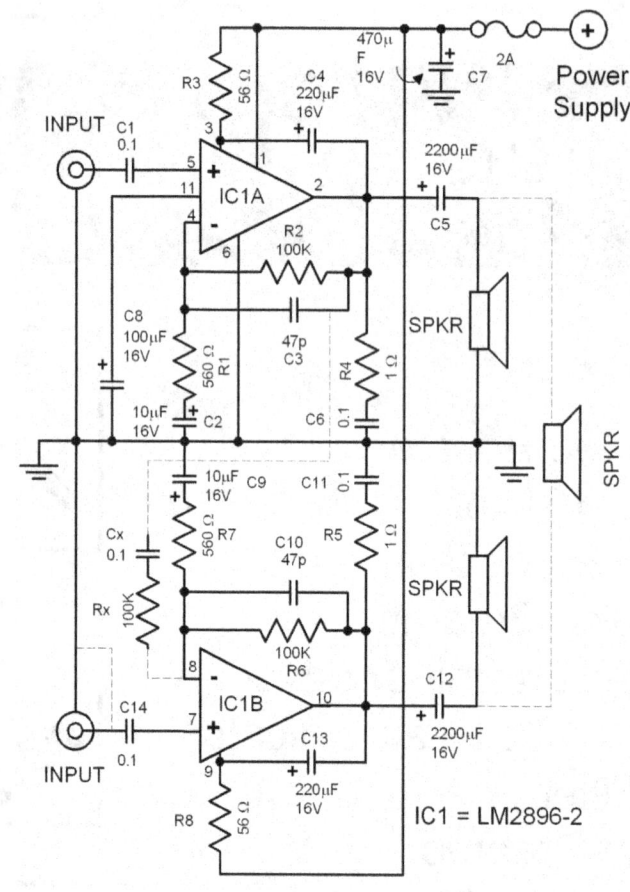

Diagram 1.0 Car Stereo Booster

This circuit uses an LM2896 IC which has two integrated amplifiers. It can be powered with voltages up to 15 volts. The power output is 2.5 watts per channel with an 8 Ω load and supply voltage of 12 volts. Using the bridge technique in the circuit gives a power output of 9 watts. The circuit can be powered with voltages from 3 up to 15 Volts. The load impedance that can be connected at its output can be either 4 Ω or 8 Ω. The supply voltage and the load impedance influence the output power level. See Table 1.0 for exact data. This amplifier circuit is designed as a booster for auto radio/cassette players. The current consumption by maximum power output and a 4 Ω load is 1 ampere.

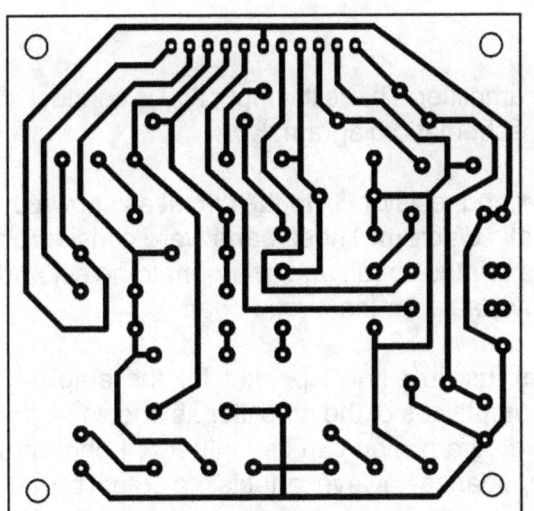

Figure 1.0 Printed Circuit Layout

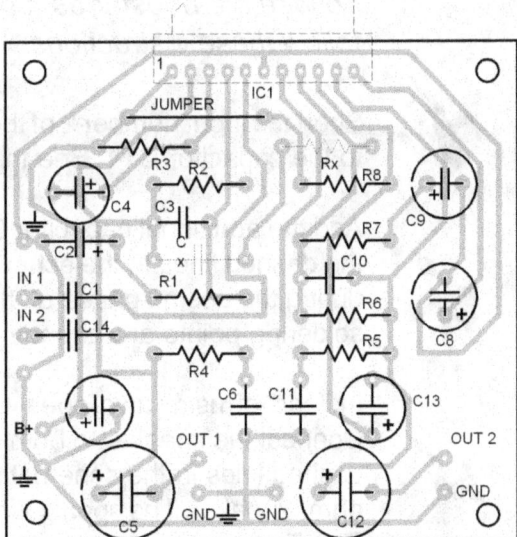

Figure 1.1 Parts Placement Layout

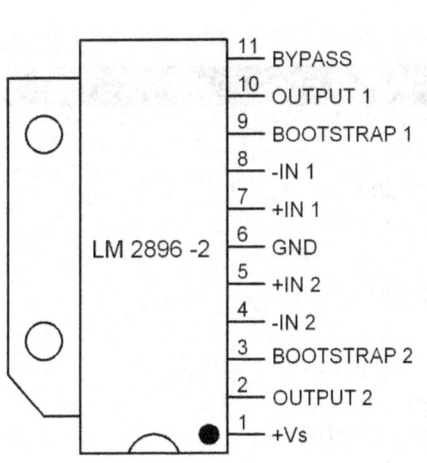

Figure 1.2 LM2896-2
Pin Designations

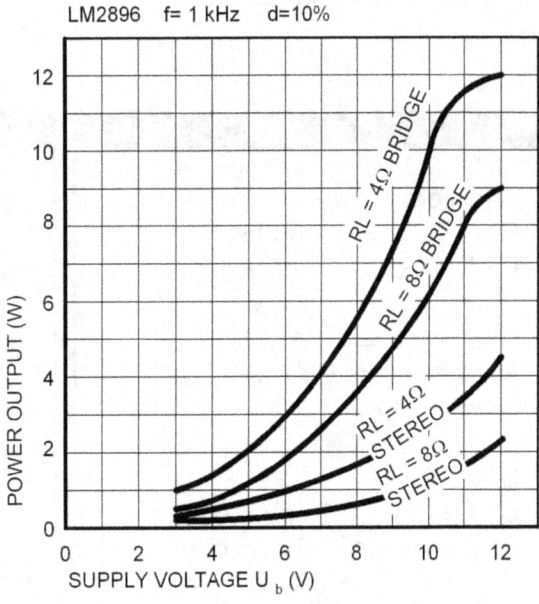

Table 1.0 Power Output vs Supply Voltage

To wire the booster as a bridged amplifier,
follow these instructions:

Short the input number 2 of the amplifier. This is the input that is connected to the capacitor C14. See the schematic diagram 1.0.

Add the additional capacitor Cx and resistor Rx to the circuit as shown by the dashed lines in the schematic diagram. These parts were considered during the design phase of the printed circuit and appropriate holes, and soldering points are already available on the circuit board.

Instead of using two speakers, use a single speaker for the amplifier. Connect the speaker to both output lines of the amplifier as shown by the dashed lines in the schematic diagram. You can use either a 4 ohm or 8 ohm speaker. The speaker impedance however affects the output power. See Table 1.0 to know what power you will get with a certain speaker impedance.

Now, since only a single speaker is connected to the amplifier circuit, you must construct a second amplifier circuit identical to this one if you want a stereo system. The second circuit will work for the second channel (and second speaker).

Parts List:	Technical Specifications:

Resistors:
R1,R7   =   560 Ω
R2,R6,Rx =   100K
R3,R8   =   56 Ω
R4,R5   =   1 Ω
All resistors are ¼ watts unless otherwise specified.
 	
Capacitors:
C1,C14,Cx= 0.1/50V Ceramic
C2,C9= 10 mF/16V Electrolytic
C3,C10= 47pF/50V Ceramic
C4,C13= 220 mF/16V Electrolytic
C5,C12= 2200 mF/16V Electrolytic
C7= 470 mF/16V Electrolytic
F1 = 2A Fuse
	Input sensitivity 20mV
	Input impedance 100KΩ
	Frequency band
	  (-3dB) 30 Hz...30kHz
	  (-3dB) bridged 30 Hz...30kHz
	
	Voltage amplification 180X
	  (bridged) 360X
	Distortion factor(f=1 kHz,Ub=12V,RL=8W)
	  -at 50mW output 0.095 %
	  -at 1 W output 0.15 %
	Power Supply 5V ... 15V
	Idle Current max. 50mA
	Output Power see Table 1.0

2 FET AUDIO MIXER

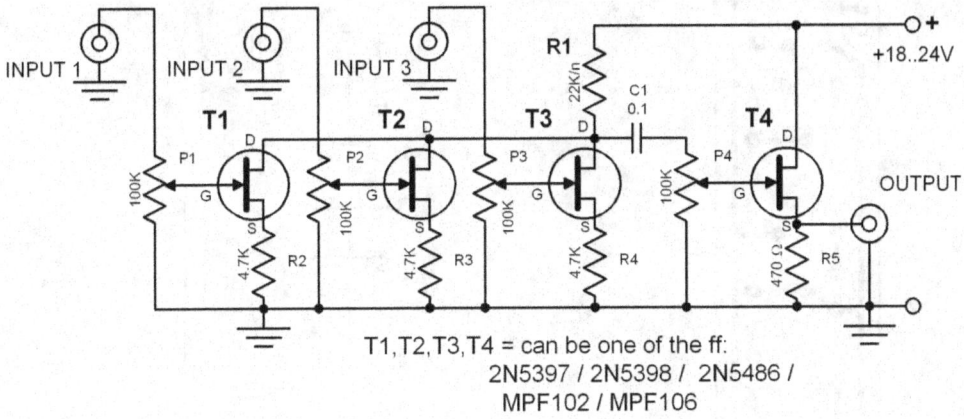

Figure 2.0 FET Audio Mixer

Although FETs are originally designed for high frequency applications they can also be used for audio frequencies. In fact, they perform excellently in this area. The audio mixer featured here is an example of the FET's versatility.

This mixer needs only four FETs and performs very well.

Formula for R1
$$R1 = \frac{22K}{n}$$
where **n** = the number of inputs

The mixer's input impedance is determined by the resistance of the input potentiometers used in the circuit since the impedance of the FETs itself is very high.

 The number of inputs that can be connected to the circuit is practically unlimited as long as the value of R1 is chosen according to the above given formula.

 The frequency response of the mixer is from 20 Hz to 80 kHz and linear within 3 dB.

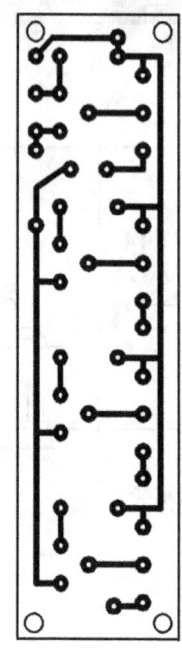

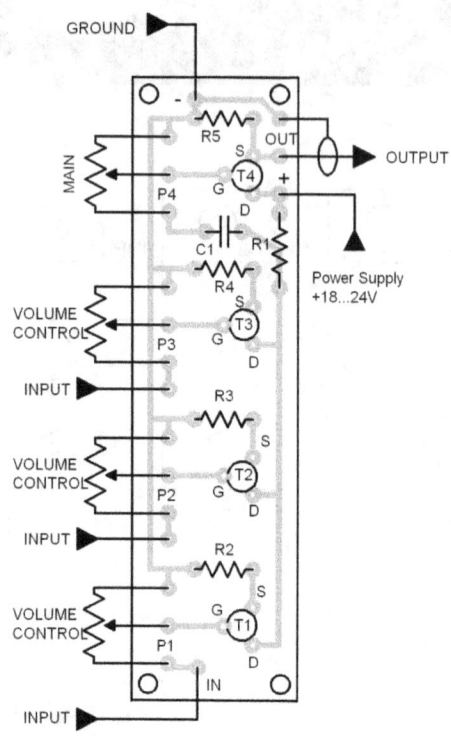

Figure 2.0 Printed Circuit Layout for the FET Mixer

Figure 2.1 Parts Placement Layout for the FET Mixer

2N5397
2N5398
MPF106

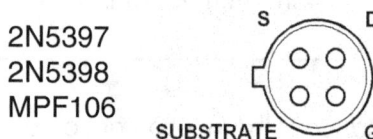

2N5486
MPF102

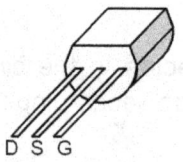

NOTE: Transistors in bottom view

Parts List:

Resistors:
R1   = 22K/n (see text)
R2,R3,R4 = 4.7K
R5   = 470 Ω
P1,P2,P3,P4 = 100K Potentiometer
All resistors are ¼ watts unless otherwise specified.

Capacitor:
C1 = 0.1µF/50V ceramic

3 SPEECH PROCESSOR

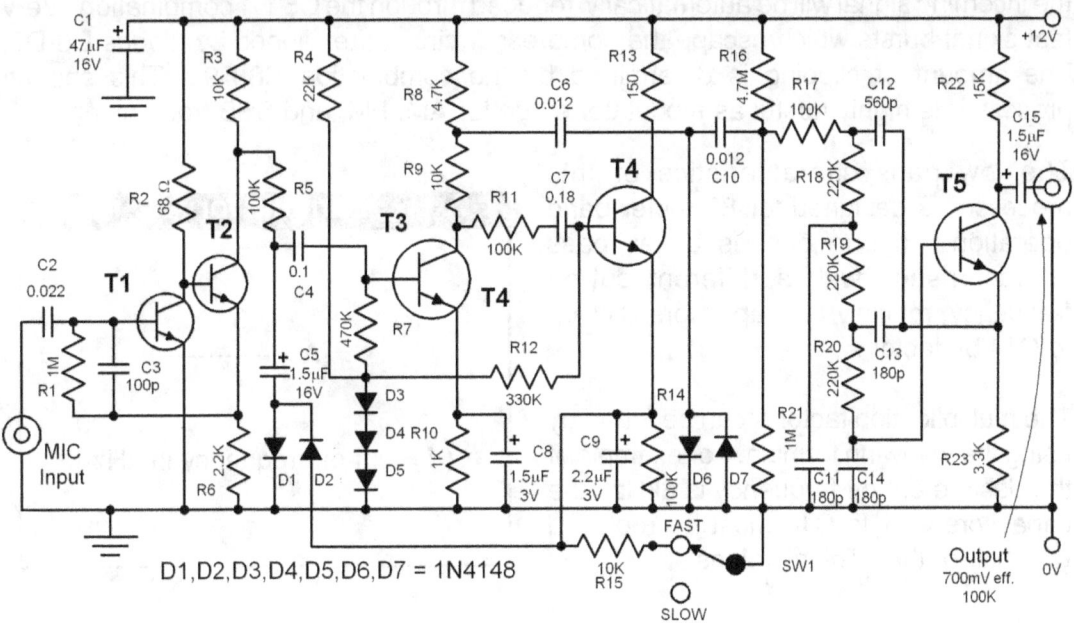

Figure 3.0 Speech Processor

There are two ways to modulate a transmitter to obtain the best possible modulation result: clipping (clipping off the amplitude peaks) and dynamic compression (generation of a constant medium powered signal carrier). Both methods have their own disadvantages. In clipping, there is no constant signal carrier, and the transmitter has a tendency to be overmodulated or undermodulated, resulting to degradation of the transmitted information. In dynamic compression, degradation also happens because of the occasional uncontrolled signal bursts due to the time constant of the compressor's regulator.

To avoid these problems, both modulation techniques are combined in one circuit. First, the modulating signal is compressed to achieve a relatively constant signal level, and then clipped to remove the unwanted amplitude peaks. The circuit first amplifies the audio signal before it is introduced into the main processing circuits. The gain of the amplifier stage (T1,T2) depends upon the impedance of the microphone being used.

This technique accomodates both low and high impedance microphones and still achieves the same signal level. Diodes D1 and D2 function together as voltage controlled signal processor. Transistor T4 on the other hand controls these two diodes.

When the voltage at the anode of D3 exceeds the threshold voltage by about 0.5 volts, the incoming signal will be automatically reduced through the C5-D1 combination. Very fast signal bursts which escape the compressor circuit are clipped by diodes D6-D7. The amount of clipping is determined by the combination R8-R9. This speech processor is highly useful as modulator stage for AM, FM, and SSB transmitters.

The low pass characteristics of this processor is designed for 80 meter band operation. If one desires a low pass characteristic with a different cut-off frequency, multiply the capacitors C11 up to C14 by factor A.

The multiplication factor A can be found by using the following formula. For example, if the desired cut-off frequency is 6kHz, the capacitors C11 to C14 must be replaced with half of the original values.

Formula for A
$$A = \frac{3}{f}$$
(**f** = cut-off frequency in kHz)

4 LOW-NOISE PREAMP

A preamplifier circuit with a very low noise characteristic can be built by simply combining a FET transistor with a normal bipolar transistor.

The input impedance of the resulting circuit is almost the same as the gate impedance of a single FET transistor - around 1 megaohm. The output impedance at the other end is about 1 kiloohm.

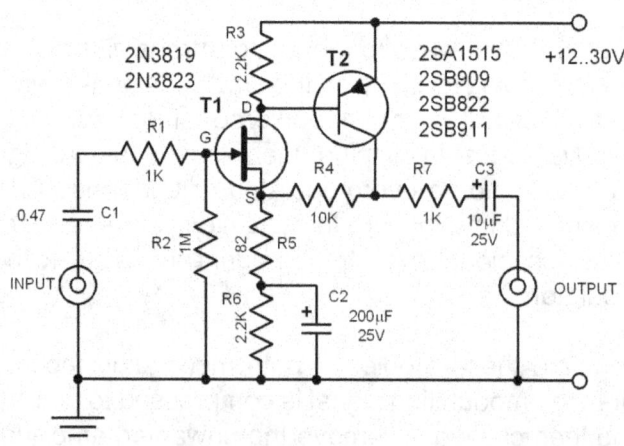

Diagram 4.0 Low Noise Preamp

The frequency of this preamplifier is linear (-3 dB) between 10 Hz and 450 kHz and (-1 dB) between 20 Hz and 200 kHz. The amplifying factor is around 100.

This circuit can be powered from 12 to 30 volts without significant deterioration in its amplifying characteristics.

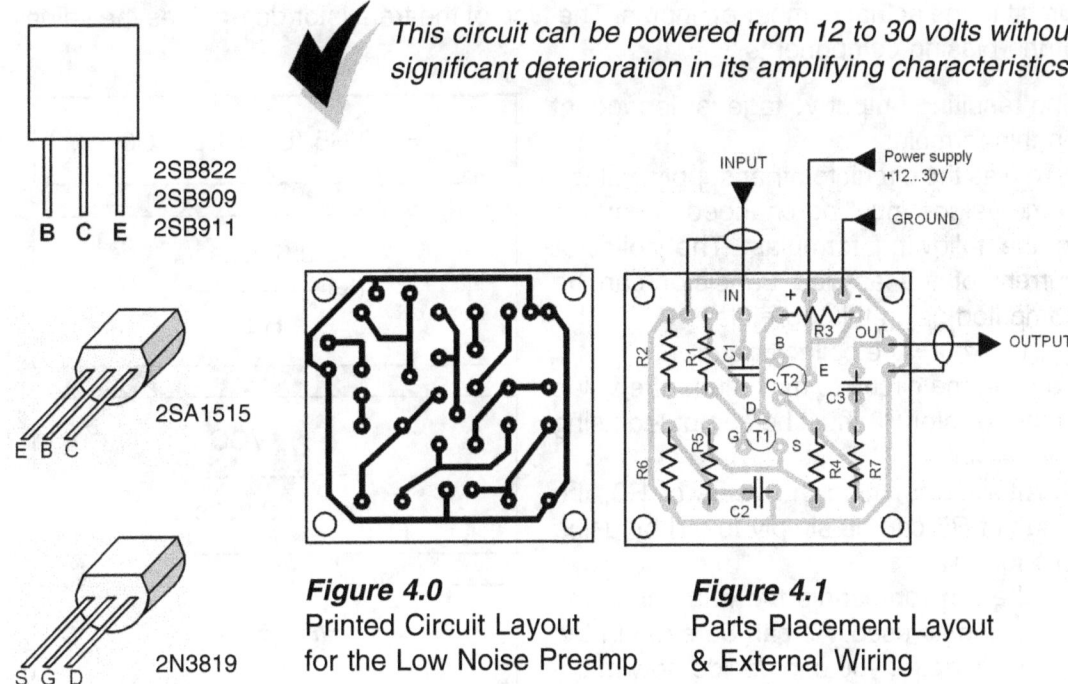

2SB822
2SB909
2SB911

B C E

2SA1515

E B C

2N3819

S G D

Figure 4.0
Printed Circuit Layout
for the Low Noise Preamp

Figure 4.1
Parts Placement Layout
& External Wiring

5 AUDIO MIXER (emitter driven)

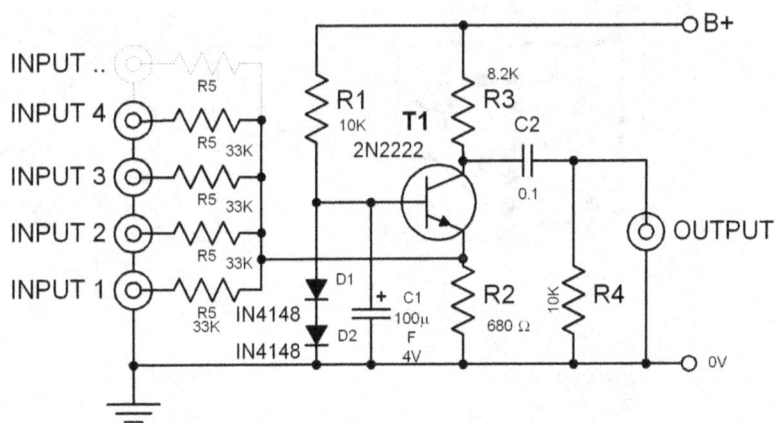

Diagram 5.0 Audio Mixer (emitter driven-single transistor)

This single transistor mixer circuit is intended for audio applications and is designed to be flexible as possible in the type of the transistor used. Additionally, the number of inputs is flexible. You simply need to change the value of a single resistor to adapt the circuit to the actual number of inputs. The type of the transistor determines the values of the biasing components

The resulting output voltage is dependent on this formula:

$$U_{output} = R3/R5\ (U1 + U2 + U3 + ...)$$

To bias the circuit to other supply voltage some values must be changed according to the following formulas: The collector current of the applied transistor can be computed using this formula:

$$i = \frac{0.6\ volts}{R2}$$

where **i** = the collector current.

To drive the circuit to maximum, the value of the resistor R3 must be computed using this formula:

$$R3 = \frac{VCC}{1.2\ volts}$$

After determining the value of R3, the value of R5 can be simply found by using this formula:

$$R5 = R3(n)$$

... where **n** represents the number of the actual inputs used. As can be seen in the circuit diagram, you can decide how many inputs you can connect as long as you follow the given formula to compute the value of resistor R5.

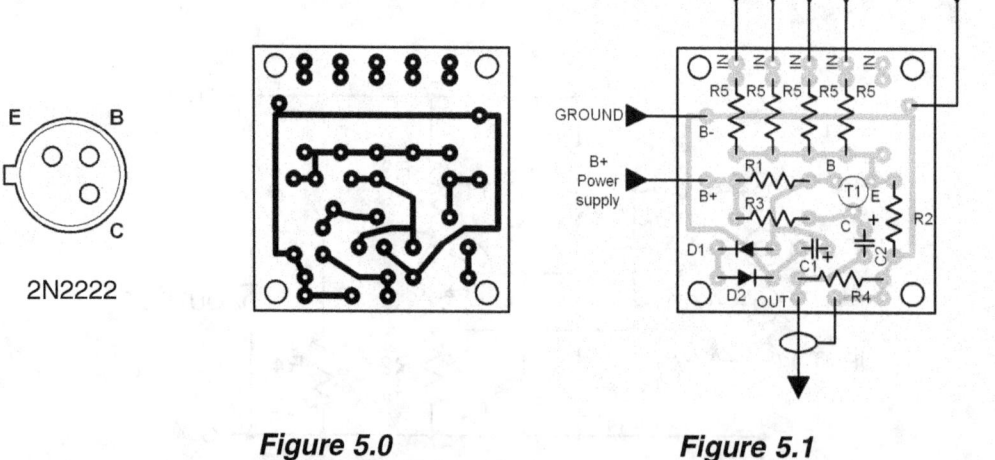

2N2222

Figure 5.0
Printed Circuit Layout
for the Audio Mixer

Figure 5.1
Parts Placement Layout
& External Wiring

6 LOUDSPEAKER PEAK INDICATOR

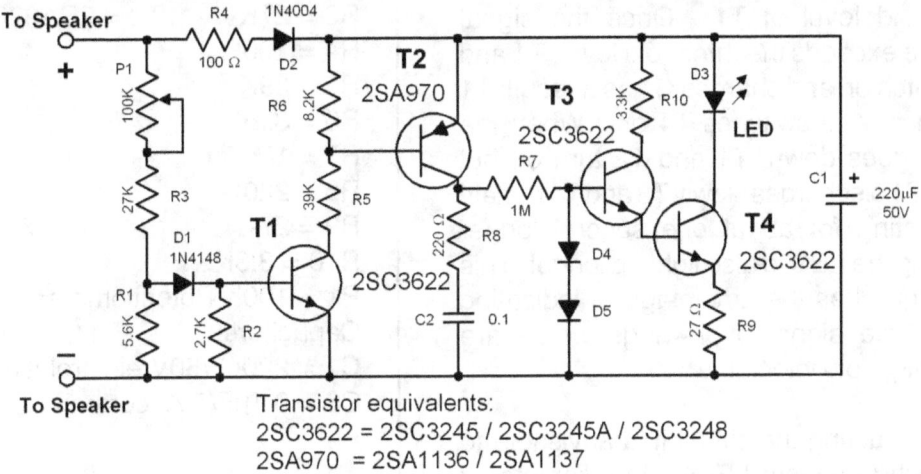

Diagram 6.0 Loudspeaker Peak Indicator

Modern loudspeaker boxes are relatively insensitive to overdriving signals, however it is still important to limit the power driving it to avoid the clipping of the audio signal. A broken sound from an overdriven loudspeaker is not only annoying to the ears but also the loadspeaker system itself can be damaged by the continous uncontrolled signal peaks.

 *Although listening can help in determining the limits of a loudspeaker system, it cannot detect short time distortions.*

This peak indicator circuit is a very useful aid in detecting the driving limits of a loudspeaker system. It can be directly connected to the existing speaker wirings and needs no extra power. It can detect very short voltage overswings and therefore provides reliable means of determining the driving limits of a loudspeaker.

The threshold level of the peak indicator can be set for speakers ranging from around 15 watts to 125 watts with 8 ohms impedance. For 4 ohms speakers, the indicator can be set for 30 watts up to 250 watts. By testing speakers, occasional blinking of the LED does not mean danger but when the LED blinks very often then it is advisable to reduce the volume of the amplifier.

The circuit operates this way: During operation the signal charges C2 through R1 and D1. In standby periods, all transistors are turned off and no current flows through the LED. A sample of the signal flows through P1 and enters T1. P1 controls the threshold level of T1. Once the signal sample exceeds the threshold level, T1 and T2 switch on and charges C1 as a result; T1 conducts and switches T4 on. When the signal goes down, T1 and T2 turn off but since C1 discharges slowly T3 and T4 remain conducting for about one second longer causing the LED to also light up longer. This technique has the advantage of indicating short time signal overswings which are normally not detectable.

In constructing the circuit it is advisable to use high luminance LEDs with a diameter of not less than 3mm.

Parts List

Resistors:	Transistors:
R1 = 5.6K	T1,T3,T4 =
R2 = 2.7K	2SC3622
R3 = 27K	T2 = 2SA970
R4 = 100Ω	
R5 = 39K	
R6 = 8.2K	
R7 = 1M	
R8 = 220Ω	
R9 = 27Ω	
R10 = 3.3K	
P1 = 100K Potentiometer	

Capacitors:
C1 = 220µF/50V electrolytic
C2 = 0.1µF/50V ceramic

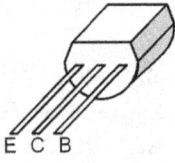

E C B

2SA970
2SA1136
2SA1137
2SC3622
2SC3245
2SC3248

Note: Transistor in bottom view

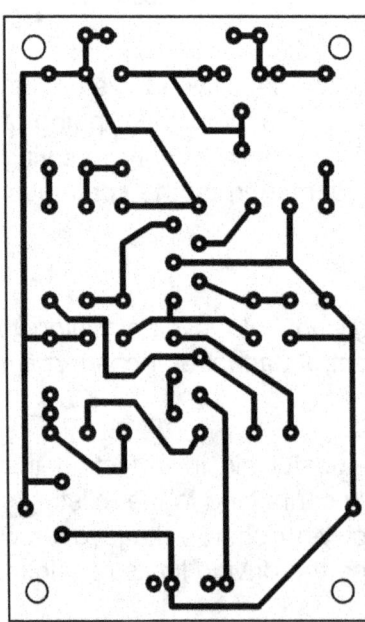

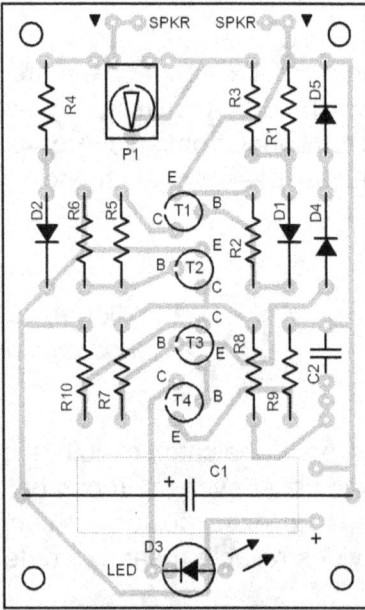

Figure 6.0
Printed Circuit Layout
for the Audio Mixer

Figure 6.1
Parts Placement Layout

7 AUDIO FILTER

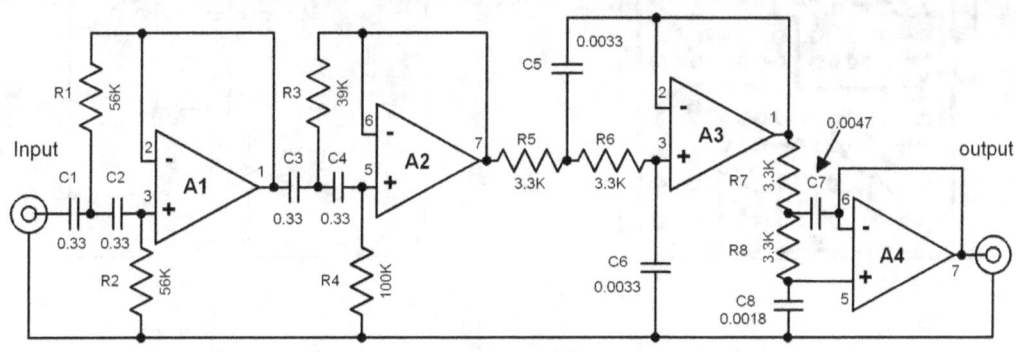

A1,A2 = IC1 = NE5532N (SE5532N)
A3,A4 = IC2 = NE5532N (SE5532N)

Diagram 7.0 Audio Filter

This is a bandpass filter for audio frequency band. It is also called noise filter. It filters unwanted signals that are lower or higher than the audio frequencies. Its bandpass characteristic is achieved through the technique of cascading a low pass filter with a high pass filter. Both filters are second-order filters with a 24 dB/octave filter capability. The 3 dB cut-off frequecies are 11.8 Hz and 10.7 kHz.

The bandpass characteristic can be changed by changing the values of the capacitors and resistors. When you want to raise the bottom cut-off frequency, you must reduce the values of C1 up to C4. On the other hand when the capacitance values are increased the bottom cut-off frequency goes down.

 If you want to reduce the top cut-off frequency, you must raise the values of R5 up to R8. If the resistance values are decreased on the other hand, the top cut-off frequency goes up.

Parts List
Resistors: R1,R2 = 56K R3 = 39K R4 = 100K R5,R6,R7,R8 = 3.3K Capacitors: C1,C2,C3,C4 = 0.33µF ceramic C5,C6 = 0.0033 µF ceramic C7 = 0.0047 µF ceramic C8 = 0.0018 µF ceramic ICs: IC1,IC2 = NE5532N

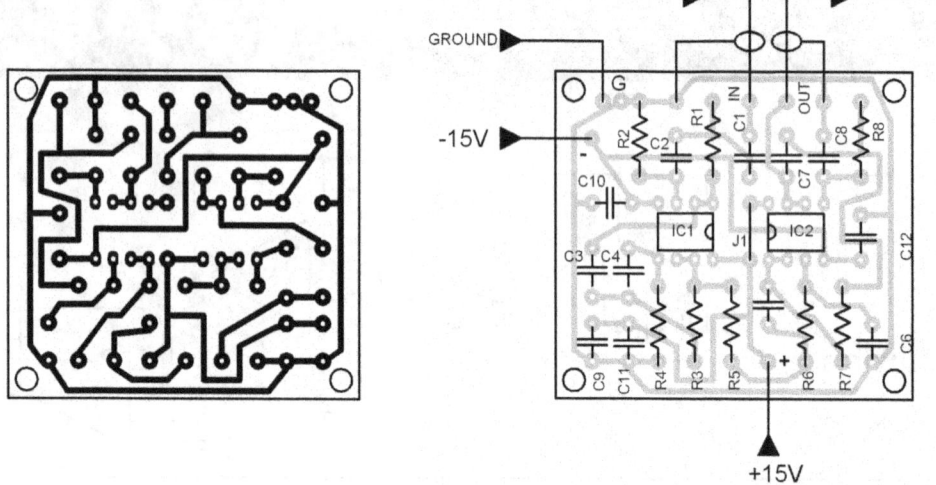

Figure 7.0 Printed Circuit Layout for the Audio Filter

Figure 7.1 Parts Placement Layout & External Wiring

8 E-GUITAR PREAMP

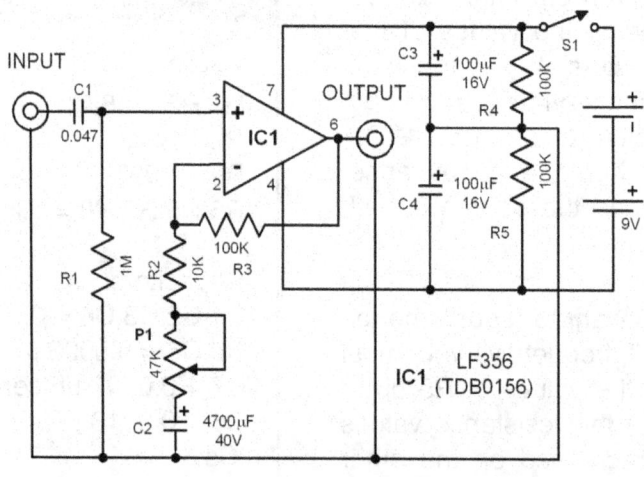

Diagram 8.0 E-Guitar Preamp

Simple but powerful. This preamp circuit is commonly used to raise the output level of an electric guitar to enable it to drive vacuum-tube amplifiers. Overdriving the vacuum-tube amplifier is sometimes necessary to achieve a certain "guitar sound". Since the normal output level of a guitar is not enough for this purpose, the small circuit featured here is just the right thing to do the „synthesizer" trick.

Technical Specifications

Amplification factor:
 -set by   R2+R3+P1/(R2+P1)
Input impedance: 1 MΩ
Supply voltage: 9 volt
Current consumption: 5 mA

This preamp amplifies the guitar output to such a level that the input stages of the final amplifier will be overdriven to their limiting level.The amplification factor of the preamp is adjustable from 3 to about 11.

As you can see clearly, the circuit is very simple. A single LF356 IC gives the signal the needed boost. The input impedance (1 MΩ) is practically determined by R1 since the IC has FET inputs.   One mega-W is an ideal value since it is the impedance of most guitar pickups. The voltage divider network R4,R5,C3 and C4 provides the IC with a symmetrical +/-4.5 volts. Due to its low current consumption, the circuit can be powered with small batteries and housed in a very compact plastic case.

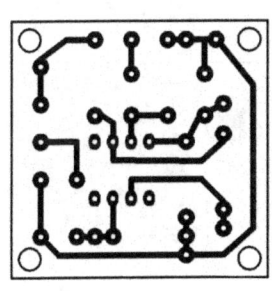

Figure 8.0
Printed Circuit Layout
for the Guitar Preamp

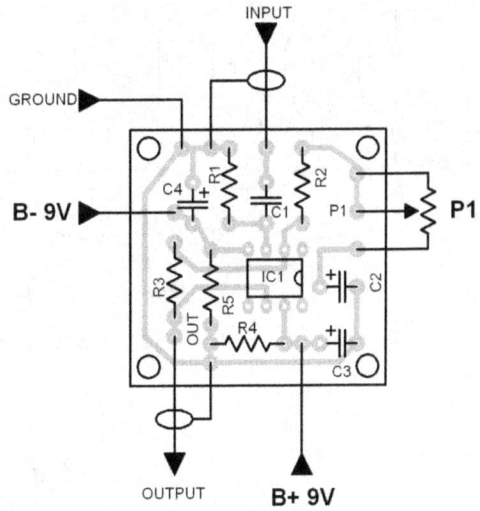

Figure 8.1
Parts Placement Layout
& External Wiring

9 40-WATTS AMPLIFIER

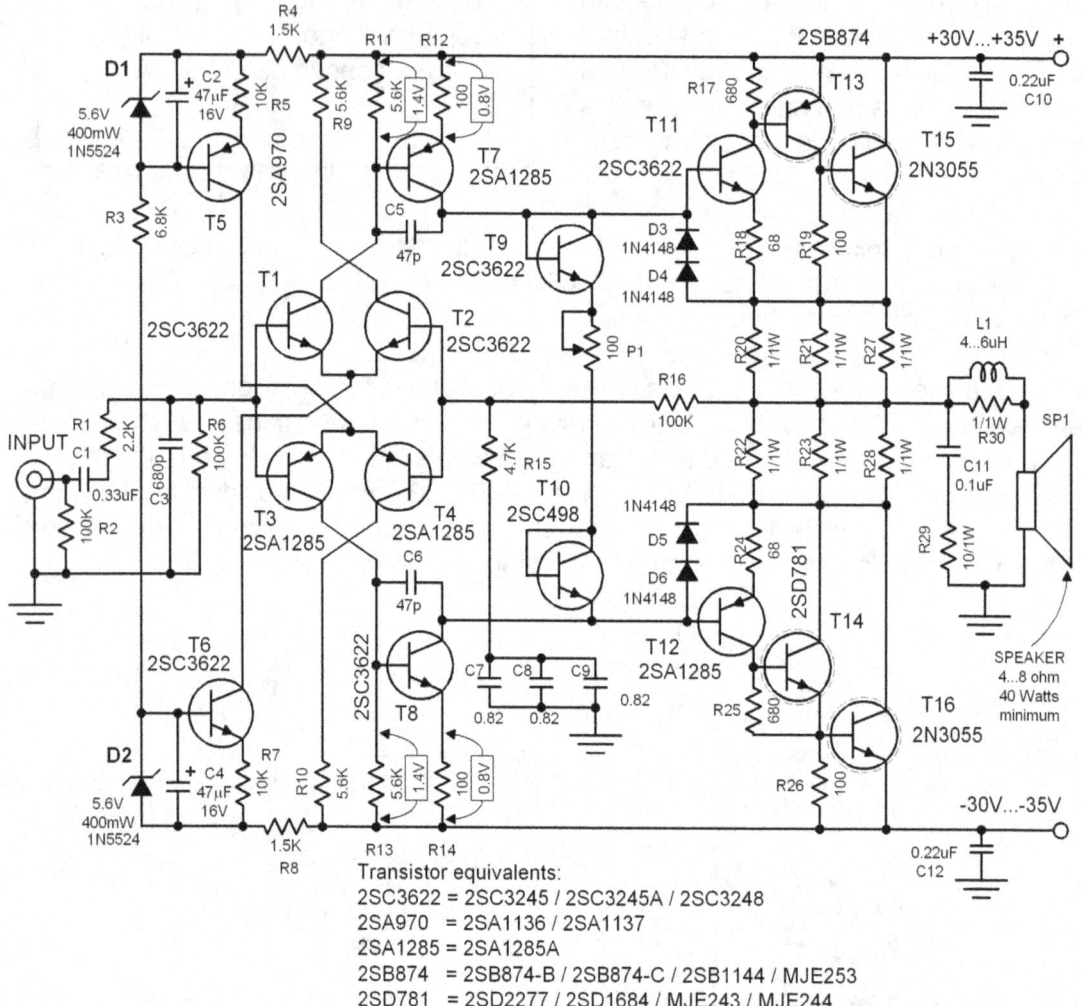

Diagram 9.0 40-Watts Amplifier

This amplifier is designed for the real music lovers who are not satisfied with just listening to the music but also want to "feel" the music. The 40 watts power from this circuit delivers a „punch". The very common 2N3055 transistors are used as final amplifiers. This makes the acquisition of the needed components easy.

At first glance, you may say that the circuit is not symmetrically designed since the final transistors are both NPN. But wait, if you will look closely at the transistor combination T12-T14-T16, you will realize that it functions as an PNP transistor. The transistor combination T11-T13-T15 on the other hand functions as NPN transistor. Since these two transistor combinations work together as a complementary pair, the circuit is symmetrical.

The rest of the circuit is designed in the same way. The transistor combination T1-T2-T6 provides a current source and functions as the upper half of a differential amplifier. The transistor combination T3-T4-T5 functions as the lower half. The transistors T7 and T8 work as drivers. The voltage drop at resistors R25...R30 is about 33 mV by 50 mA standby current.

Transistors T15 and T16 must be heatsinked. Be sure that the transistors are electrically isolated from the heatsinks. Transistors T13 and T14 must also be heatsinked. Transistor pair T9 and T11 must be thermally coupled, meaning they must have a common heatsink. Transistors T10 and T12 must also be thermally coupled.

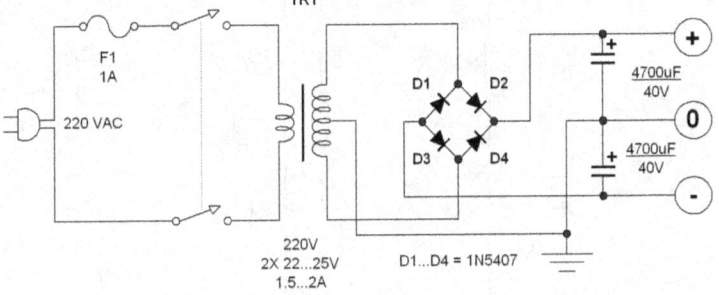

Diagram 9.1 Power Supply (40-Watts Amplifier)

Coil L1 must be self constructed: Wind 10 turns of magnet wire around R7. The ends of the magnet wire must be soldered to the terminals of R7.

TAKE NOTE

Only one channel of the circuit is shown. To build a stereo system, you must build two identical circuits.

Technical Specifications	
Distortion factor:	0.01 % in 20 Hz - 20 kHz
Power output:	40 Watts (8 Ohms);
60 Watts (4 ohms)	
Input sensitivity:	800 mV for 40 W
	850 mV for 45 W
	700 mV for 60 W
	725 mV for 65 W
Frequency Response:	15 Hz - 100 kHz (+/- 1 dB)
Current amplification:	approx. 200,000
Standby current:	25 - 50 mA
Current consumption:	1 A by 40 W
	1.06 A by 45 W
	1.75 A by 60 W
	1.81 A by 65 W
** W = watts	

10 AUDIO WATTMETER

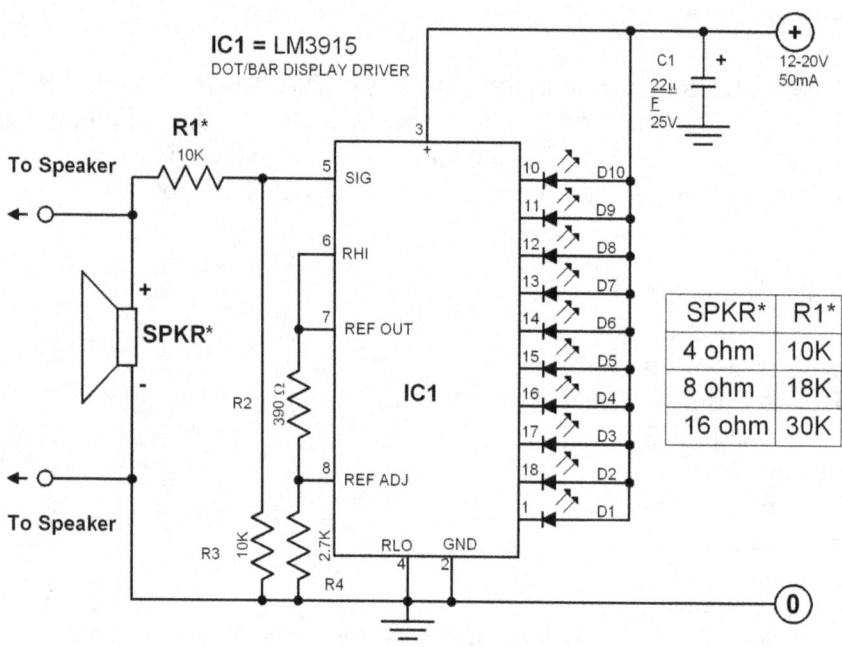

Diagram 10.0 Audio Wattmeter

You want to measure the power output of your audio amplifier? Well, take one LM3915 IC and add a few passive components .. then viola! You've got a simple but effective audio wattmeter.

This circuit uses a row of colored LEDs as a scale to show the relative power output of your amplifier in watts. It is easy to be inserted into the speaker box. All you need is to hook a supply voltage to it. The value of R1 depends upon the impedance of the speaker being used. The small table near the diagram shows the necessary values of R1.

 *If you want to apply the circuit to stereo systems you must build two identical circuits.*

The supply voltage of the circuit is 12...20 volts/50 mA DC adapter. Take note that the LM3915 indicates only the positive swing of the signal. Anyway, by testing amplifiers it does not matter anyhow since a sine wave is normally used as a test signal. To test the circuit without using an actual speaker, you can connect a dummy resistor with a value of either 8 or 4 ohms.

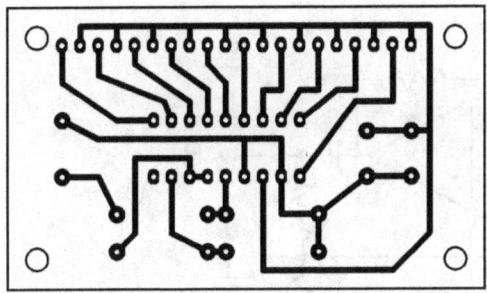

Figure 10.0
Printed Circuit Layout for
the Audio Wattmeter

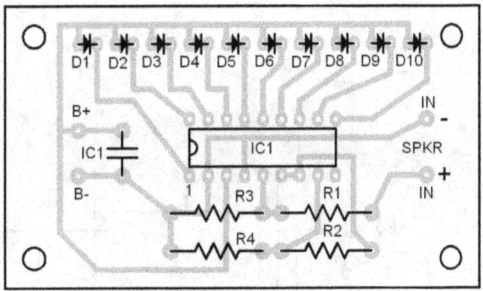

Figure 10.1
Parts Placement Layout

11 ELECTRONIC SWITCHBOARD

You know the problem only too well: signal sources of all kinds, cable salad, connectors. Well, if electronics brought this problem why not use electronics to solve it as well. Right! ... and the solution is an electronic signal switchboard.

This switchboard circuit is modifiable according to one's needs. With this circuit you can say goodbye to cable salad.

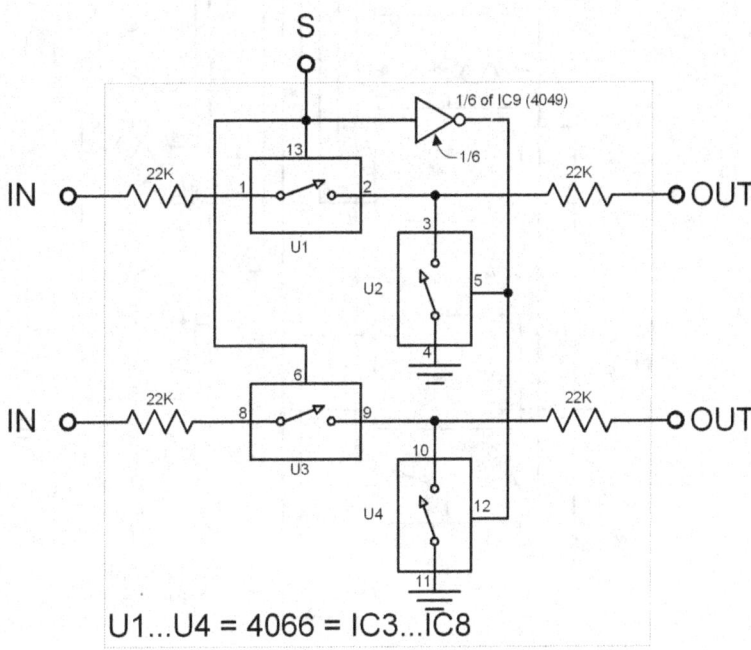

Diagram 11.0 Electronic Switchboard

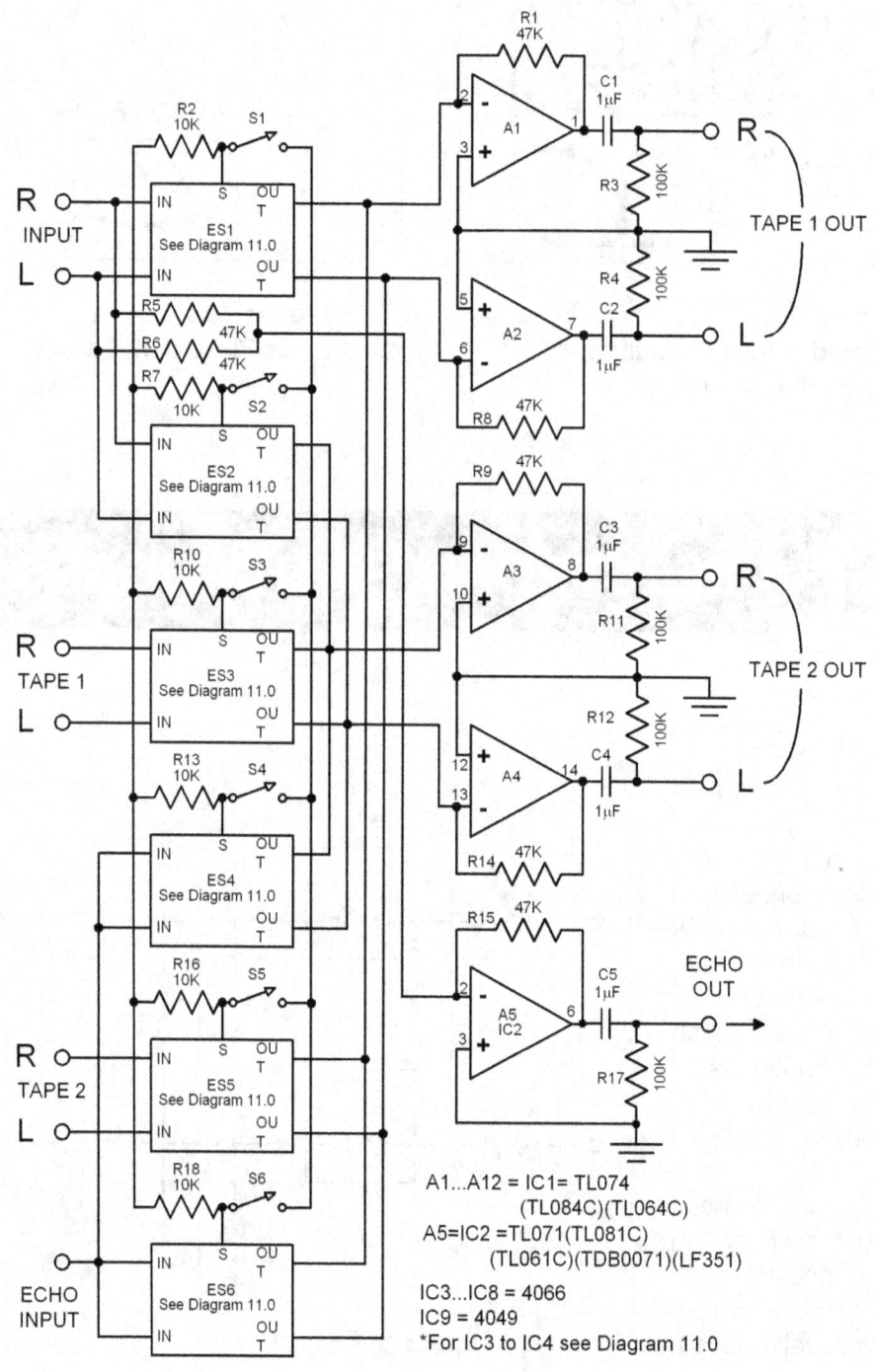

Diagram 11.1 Electronic Switchboard

The circuit featured here can be easily adapted to your needs. It has one free definable input, two tape recorder inputs and one echo effect input. All four inputs can be connected to any of the three outputs (two tape recorder outputs and one echo effect output). The changing of the connections is done by simply flipping any of the six toggle switches which controls the electronic analog switches.

The heart of the circuit is composed of 6 electronic analog switch modules (ES1 up to ES6) which you see on the left part of Diagram 11.1. Detailed circuit design is shown in diagram 10.0. One ES module uses one 4066 analog switch and 1/6 of the 4049 inverter of IC9.

Technical Specifications	
Channel separation:	75 dB
Signal/noise ratio:	more than 100 dB
Distortion factor:	less than 0.01 %
Current consumption:	approx. 20 mA
Supply voltage:	two 9 Volt blocks

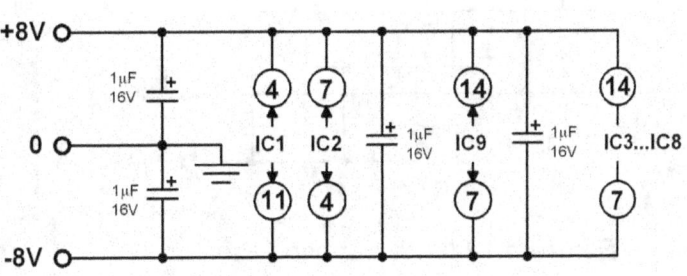

Diagram 11.2 Supply connection to IC's

The inner function of each of these switches is shown in Diagram 11.0. The inner switches work much like ordinary relays. They are controlled by a voltage that is applied to the S-input. Each signal channel needs two electronic switches. When the voltage applied to S-input is positive, all the inner switches will open, preventing the input signal to get through. When the S-input sees a negative voltage, the inner switches will close.The output signal is finally amplified by IC5. The amplification factor is variable through R11.

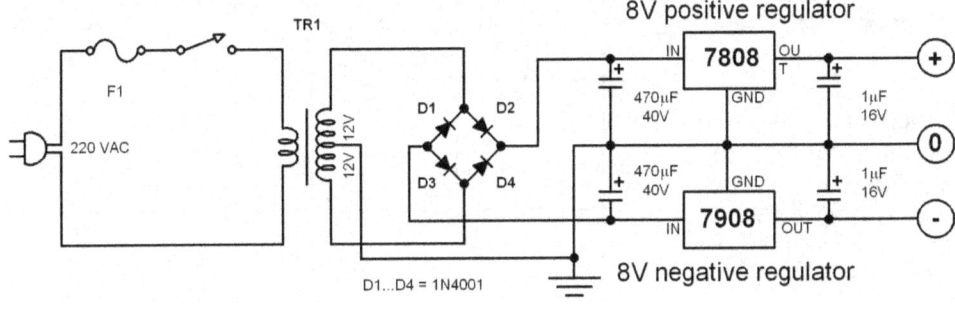

Diagram 11.3 Power Supply (Electronic Switchboard)

12 MICROPHONE PREAMP

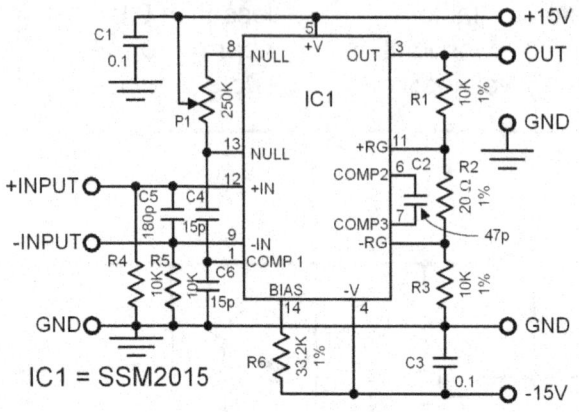

IC1 = SSM2015

This microphone preamp is extremely quite exhibiting an extremely low noise factor of 1.3 nV/sqr(Hz). Additionally, it has a high voltage amplification factor between 10 and 2000. This factor (G) comes from the simplified formula:

$$G = (2000/R2) + 3.5$$

$$\text{when } R1 = R3 = 10 \text{ K}\Omega$$

Diagram 12.0 Microphone Preamp

In the featured application circuit, the amplification factor is set to 1000. Resistor R6 affects the bandwidth and slew rate. The value of 33k for R6 provides a good compromise. This value, however, causes a standby current of 4.5µA. Anyway, we can achieve a noise factor of 95 dB with the inputs short circuited or 86 dB with an input impedance of 600 ohms. If the value of R6 is changed, the value of C4 and C6 must also be changed. Table 12.0 shows the necessary values.

The differential inputs of SSM2015 "float" that is why resistors R1 and R2 are added. These two resistors stabilize the DC current input. If the chip is used for so-called single-ended-applications, it is necessary to adjust P1 to compensate for input offsets. See Table 12.1.

Potentiometer P1 does not affect the amplification factor of the IC. The IC can drive capacitive loads up to 150pF. It has a bandwidth of 180 kHz (-3dB). The output voltage is 3 volts with a load of 1 kiloohm.

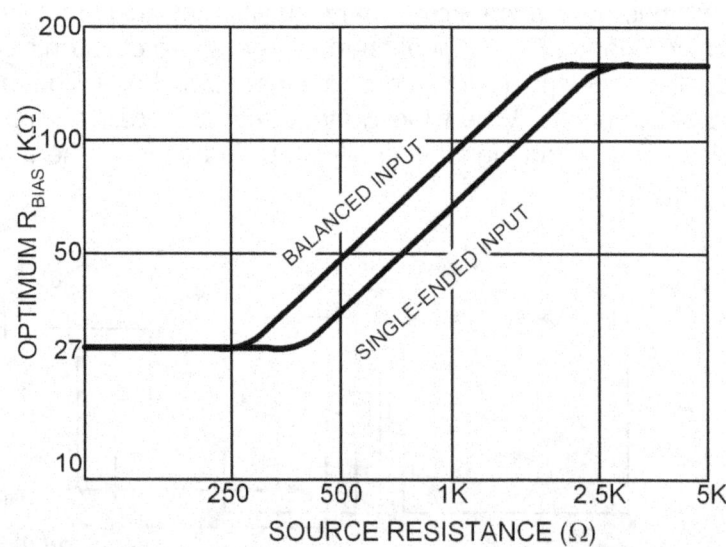

Graph 12.0 Source to Bias Resistance

Table 12.0		
R6	**C4**	**C6**
27K...47K	15p	15p
47K...68K	10p	15p
68K...150K	5p	20p

G	R6 = 27K..47K	47..68K	68K..150K
		Table 12.1	
10	500K	250K	250K
100	500K	100K	100K
100	250K	100K	50K

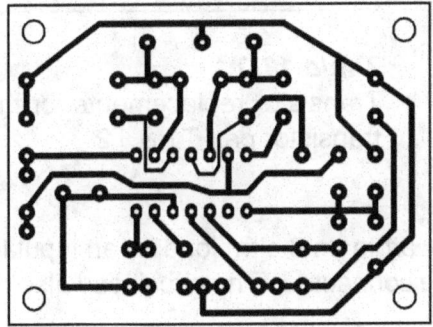

Figure 12.0 Printed Circuit Layout for the Microphone Preamp

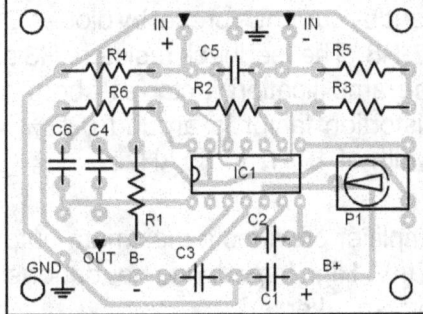

Figure 10.1 Parts Placement Layout for the Microphone Preamp

13 HIFI HEADPHONE AMPLIFIER

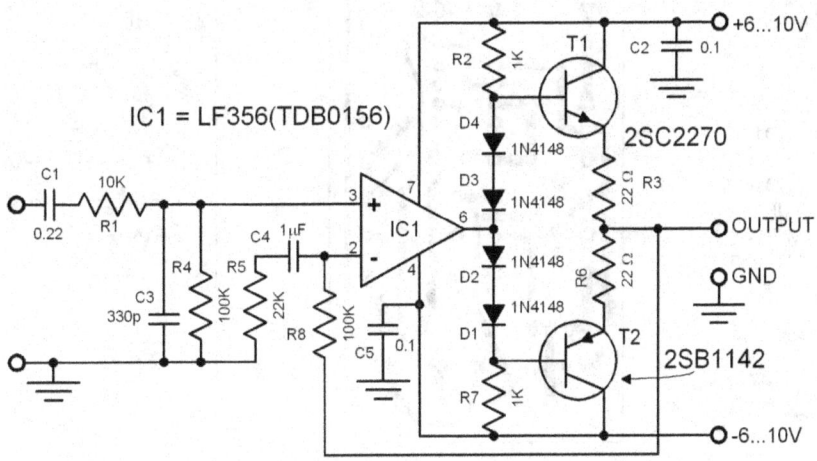

IC1 = LF356(TDB0156)

Diagram 13.0 HIFI Headphone Amplifier

This amplifier is normally used to drive a headphone with a relatively low impedance. It provides 1 watt power output. It can also be applied as an output stage for preamplifiers in conjunction with active loudspeaker boxes. The circuit is composed of an OP-AMP and an additional transistor amplifier. Input signals pass through a low pass filter composed of R1-C2. Its application together with a relatively "fast" OPAMP provides a low distortion factor. The standby current is preset by diodes D1...D4 and R7-R8. The feedback resistors R3 and R4 set the amplification factor at about 15 dB. The distortion factor is around 0.1 % with a bandwidth of 10 Hz to 30 kHz.

PAIR 1	T1 One of the ff:	T2 One of the ff:
$U_c = 45\ V_{max}$ $I_c = 1.5\ A_{max}$	2SC3420 2SD826 2SD1685 2N6412	2SB1143 2N6414

PAIR 2	T1 One of the ff:	T2 One of the ff:
$U_c = 100\ V_{max}$ $I_c = 1.5\ A_{max}$	2SD781 2SD1177 2SD1684 MJE243 MJE244	2SB874 2SB874.B 2SB874-C 2SB1144 MJE253

Table 13.0
Transistor replacements for the transistor pair T1 & T2

The amplifier provides a maximum output of 1 watt to an 8 ohm load by an input level of 500 mV. High impedance headphones can be connected. The circuit can also drive a 4 ohm speaker.

In order to prevent the output transistors from being destroyed in case of an output short circuit, they must be heatsinked. The transistors must also be electrically isolated from the heatsink. Supply voltage can be provided by an adaptor with 6-8 volts DC output.

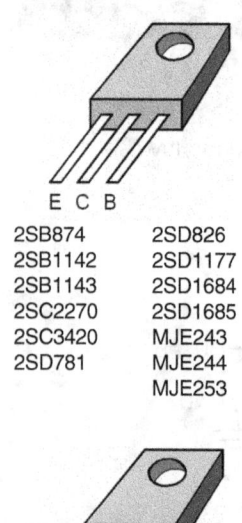

E C B

2SB874   2SD826
2SB1142   2SD1177
2SB1143   2SD1684
2SC2270   2SD1685
2SC3420   MJE243
2SD781   MJE244
          MJE253

B C E   2N6412
         2N6414

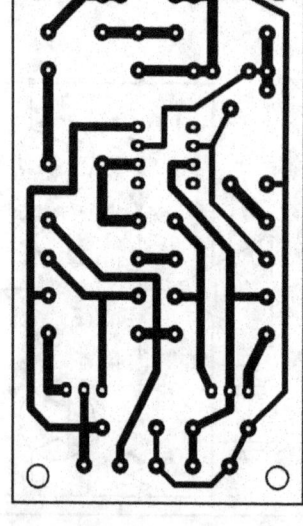

Figure 13.1
Printed Circuit Layout

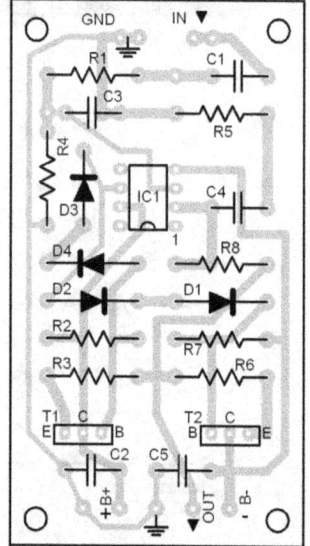

Figure 13.1
Parts Placement Layout

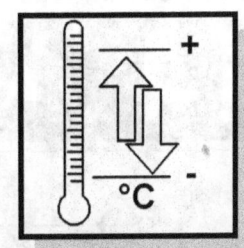

CONTROLLERS & SENSORS

36 Automatic Light Dimmer
37 Motor Speed Regulator
38 Presence Detector
39 Drill-Speed Regulator
41 Extended-Time Timer
41 Motor Speed Control
44 Intercom
45 Temperature Alarm
46 Electronic Dog Whistle
47 Coffee Thermometer
49 Telephone Lamp Control
50 °C Thermo-Sensor

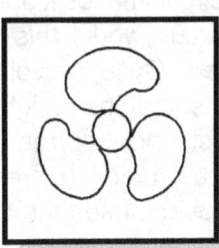

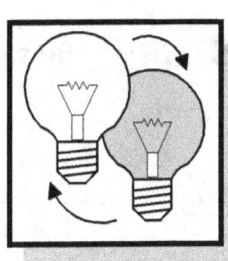

14 AUTOMATIC LIGHT DIMMER

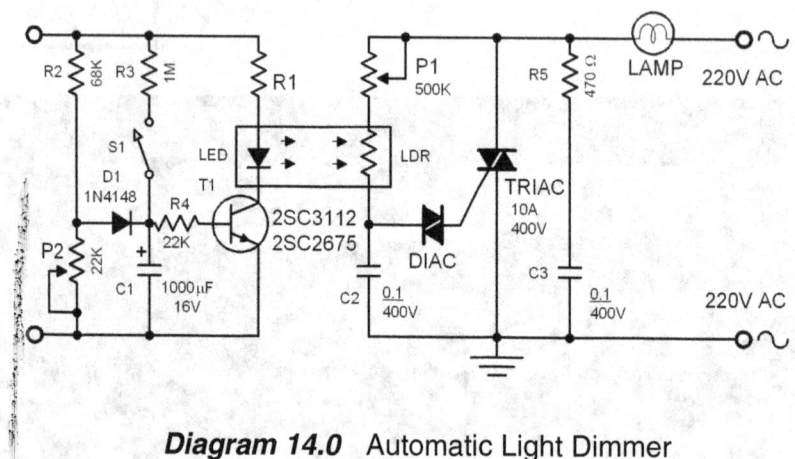

Diagram 14.0 Automatic Light Dimmer

This circuit makes it possible to control a lighting system so that it turns on or off slowly. The circuit works this way: When switch S1 is closed, the capacitor C1 is slowly charged. Once the voltage at C1 reaches 0.6, transistor T1 begins to conduct and the LED also begins to light. If the capacitor voltage increases further, then transistor T1 conducts more current and in return the LED lights brighter. If the LED lights up, the LDR resistance decreases causing the SCR to conduct periodically earlier. This technique causes the lighting system to turn on slowly.

On the other hand if switch S1 is turned off, meaning the switch is opened, the LED does not immediately turn off since the capacitor voltage at the base of T1 discharges slowly. The LED slowly dims until it finally turns off. This causes the lighting system to slowly dim out before it finally turns off. Potentiometer P2 must be set so that the anode voltage of D1 is about 0.7 volts. If this is done, the capacitor voltage will be around 0.5 volts during standby, meaning lights off.

Never use replacement diode types for diode D1.
Be sure to use an originally marked 1N4148 signal
switching diode.

Caution! Danger of electrocution!
You are working with a line voltage of 220 Volts AC.
Extreme shock hazard!

15 MOTOR SPEED REGULATOR

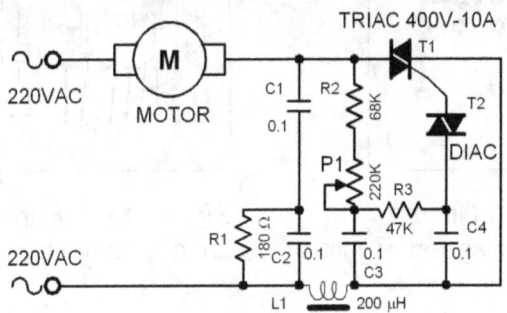

Diagram 15.0 Motor Speed Regulator

This triac-based speed regulator is designed for controlling the speed of small household motors like drill machines. The speed of the motor (example: drill motor) can be controlled by changing the setting of P1. The setting of P1 determines the phase of the trigger pulse that fires the triac. The circuit incorporates a self-stabilizing technique that maintains the speed of the motor even when it is loaded. For example:

When the motor of the drill machine is slowed down by the resistance of the drilled object, the counter-EMF of the motor also decreases. This results to a voltage increase in R2-P1 and C3 causing the triac to be triggered earlier, and the speed increases accordingly.

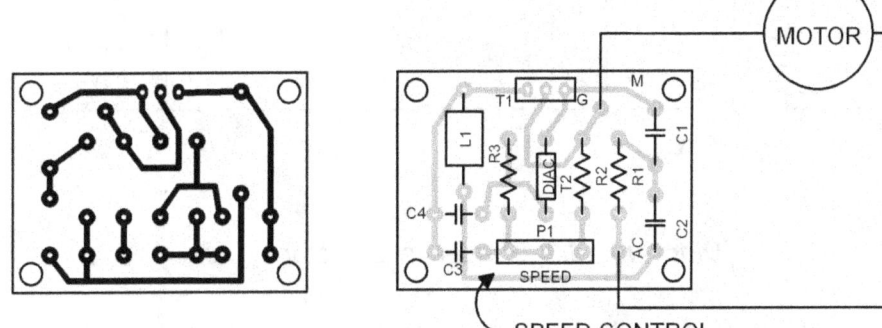

Figure 15.0 Printed Circuit
Layout #1

Figure 15.1 Parts Placement Layout #1

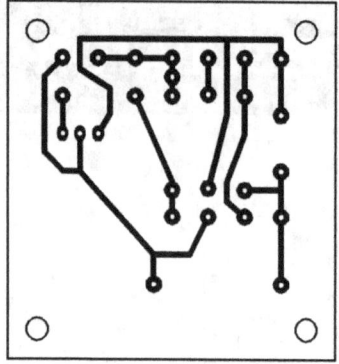

Figure 15.2 Printed Circuit
Layout #2 for motor speed regulator

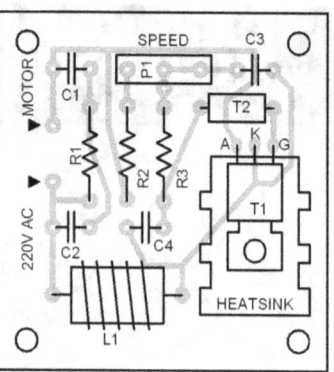

Figure 15.3 Parts Placement
Layout #2 for motor speed regulator

16 PRESENCE DETECTOR

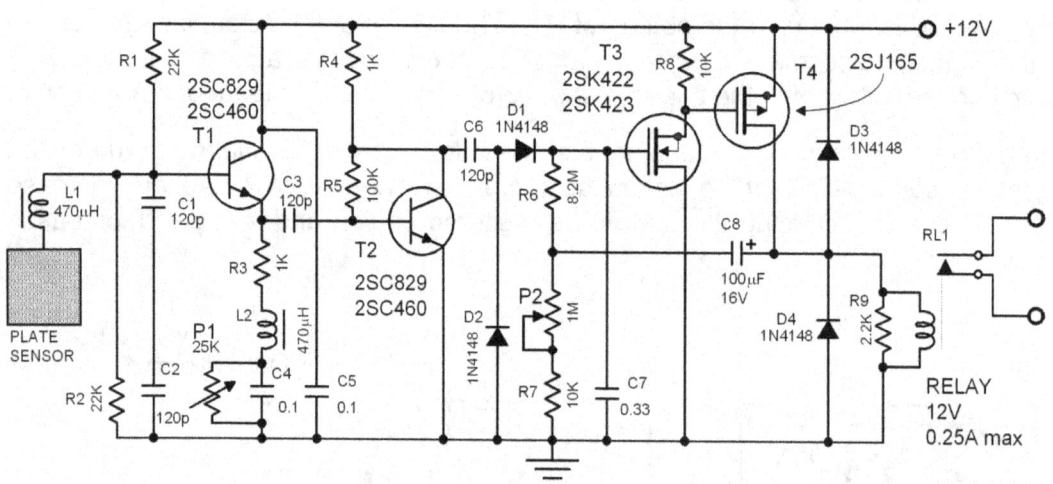

Diagram 16.0 Presence Detector

This detector circuits reacts to a conductive object within a certain defined area. For example if somebody walks near the sensor plate of the detector circuit, the relay closes thereby switching on whatever signalling device is connected to it. This circuit , however, does not register the actual movement of the conductive object. It only detects the presence of the object.

The sensitivity of the detector can be varied through P1 so that the chosen "triggering distance" can be set as needed. One possible application of this circuit is the automatic control of doors and gates. Another is detection of metal pipes inside concrete walls. The circuit can be applied as level detector of corrosive liquids in glass containers since the sensor plate does not touch the liquid. It can also be used as a universal touch switch, etc.

The circuit is primarily a very stable Clapp oscillator. The sensor plate works as an unstable capacitor with capacitance lower than C1 and C2. The featured circuit oscillates at approximately 1 kHz. The oscillator signal is amplified and then rectified. The rectified signal serves as a control pulse to trigger the monostable circuit T3 and T4. The use of VFETS as monostable components alllows it to directly control the relay. The triggering time is set by P2.

During operation make sure that the sensor plate is placed far enough from metallic objects except when the circuit is used as a metal detector. Otherwise the capacitance sensed by the oscillator becomes very large and the relay will stay closed. The detector can be controlled remotely by placing the sensitivity control P1 away from the circuit.

17 DRILL SPEED REGULATOR

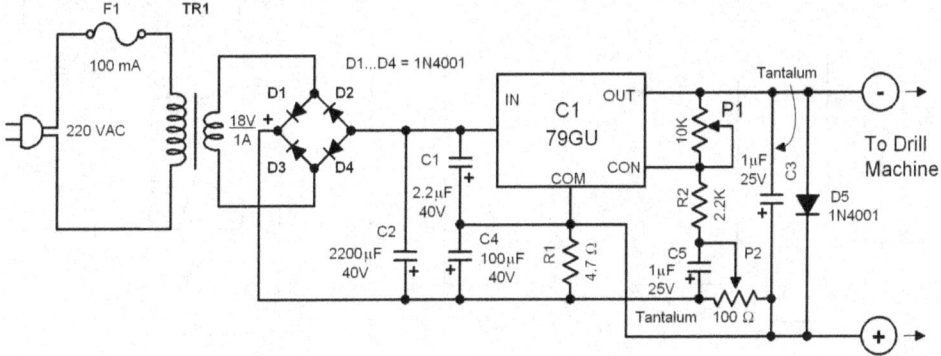

Diagram 17.0 Drill Speed Regulator

A mini-drill machine is always needed in drilling the printed circuit board. It is usually powered by batteries that limit the drill's capacity. Using a power supply adaptor with a speed regulator like the one featured here can certainly improve the performance of the drill. The speed regulator can maintain the speed of the drill no matter what load is encountered by the drill.

Potentiometer P2 controls the speed of the mini-drill. Potentiometer P1 sets the maximum voltage allowed to power the mini-drill.

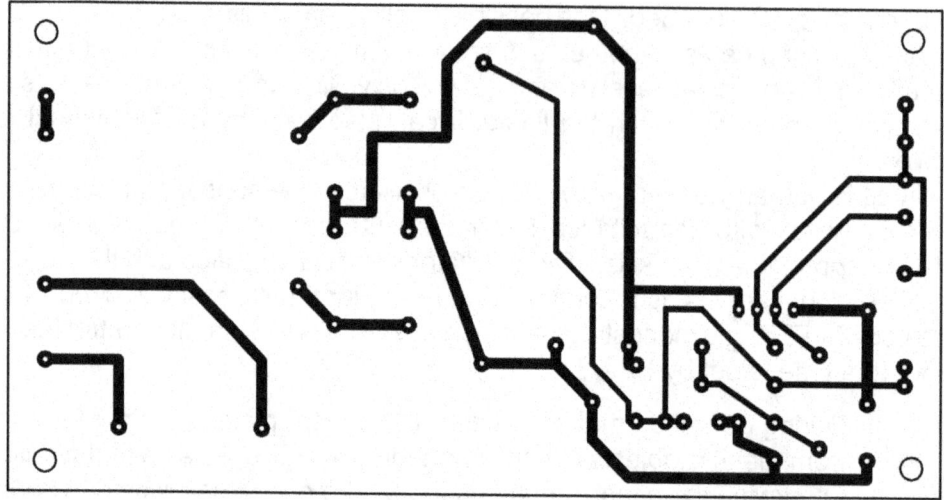

Figure 17.0 Printed Circuit Layout for the Drill Speed Regulator

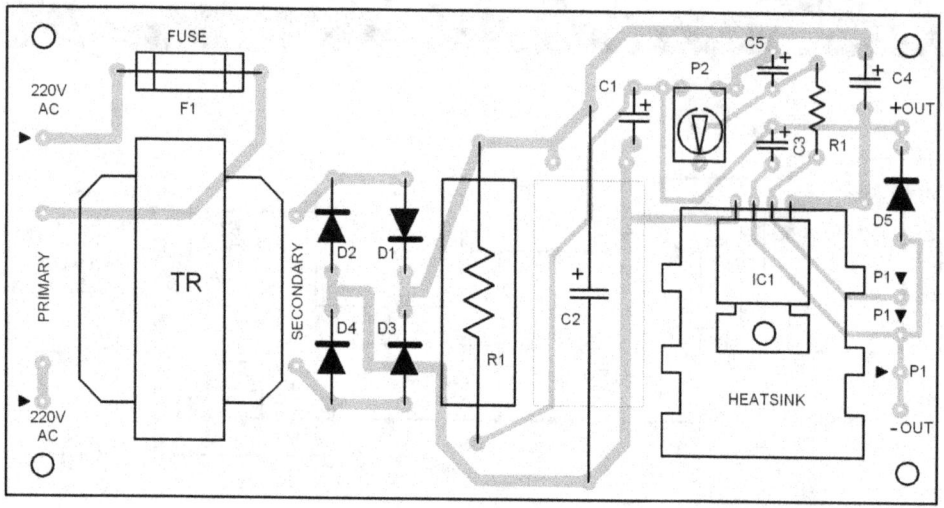

Figure 17.1 Parts Placement Layout for the Drill Speed Regulator

Caution! Danger of electrocution! Extreme shock hazard! You are working with a line voltage of 220 Volts AC.

18 EXTENDED TIME TIMER

Extended-time timers which range in minutes are usually designed with very high R-C value components. This often leads to complicated circuit designs. With a bit of creativity, however, an extended-time timer circuit with simple low R-C value components can be constructed.

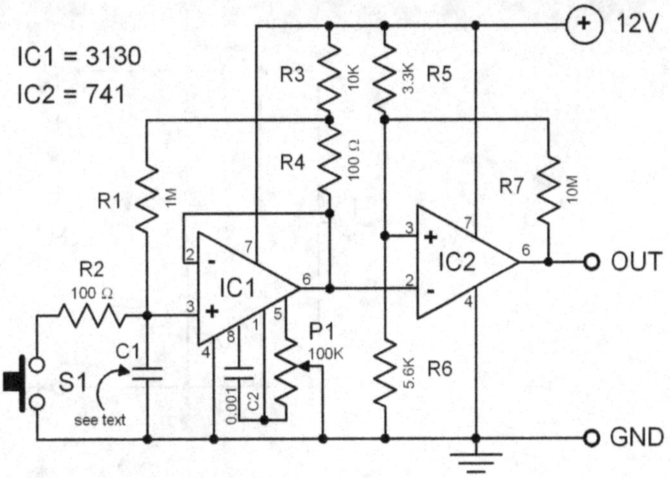

Figure 18.0 Extended Time Timer

The featured circuit functions this way:

The opamp IC1 prevents the capacitor C1 to discharge immediately. Without the IC1 the capacitor C1 would be immediately discharged through R2. The voltage drop at R1 is almost equal to supply voltage when the capacitor is fully charged. IC1 functions as a current follower and its output voltage is equal to the voltage at capacitor C1.

The threshold voltage at the minus input of IC2 is preset by R5-R67. Once the output voltage of IC1 reaches the threshold, IC2 sends out a voltage pulse. Resistor R78 is chosen so that the hysteresis is very small causing IC2 to produce only a pulse. The timer is started by discharging the capacitor C1 through switch S1. Potentiometer P1 must be adjusted in such a way that the circuit will stabilize after switch S1 is released. The switching time of the circuit is independent from the supply voltage. It works accurately with supply voltages from 10 to 15 volts. With supply voltages higher than 15 volts, the circuit will become unstable.

Switching time T can be computed using the given formula:

The component values of the featured circuit gives a switching time of: $T = 100R_1C_1$ If for example C1=1µF then: T= 100 seconds. To enable easy variation of the switching time, replace resistor R1 with a potentiometer.

Formula for T

$$T = R_1C_1\left(1 + \frac{R_3}{R_4} + \frac{R_3}{R_1}\right)\ln\left(1 + \frac{R_6}{R_5}\right)$$

19 MOTOR SPEED CONTROL

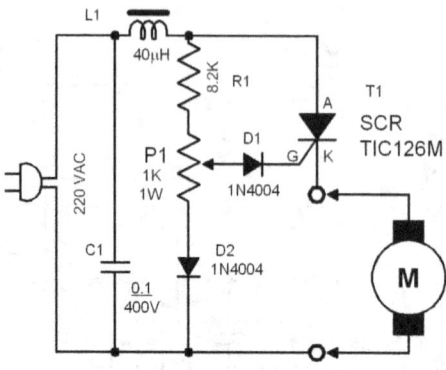

Figure 19.0   Motor Speed Control

Only seven components regulate the speed of a motor! Simple and very cheap to construct. This circuit controls the speed of a motor from 0 up to 50 % of its nominal power. Resistor R1 and potentiometer P1 divide the positive halfwave swing of the line voltage. During negative swings, the current is blocked by diodes D1 and D2 and the thyristor receives no triggering voltage. The thyristor, in turn, cannot conduct current.

But why can the circuit control only up to 50 % of the maximal motor power? Well, speed regulation is only practical at low motor speeds. At more than 50 % of the motor's capacity, speed regulation becomes almost unnecessary. Anyway, below 50%, this circuit has the capability to precisely set the motor's speed.

If you want to be able to control the current up to 100 % like in applying this circuit as lamp dimmer, you only need to connect a bridge rectifier between the line voltage and the circuit. In this case the bridge rectifier delivers both voltage swings into the circuit.   You must choose a bridge rectifier with correct values

 Four 1N4004 are enough to handle lamps up to 200 watts. If you want to handle lamps up to 600 watts, you'd better use four 1N5404.

Parts List:
P1= 1K/1W with plastic handle
R1= 8.2 Ω/5 w
C1= 0.1µF/250VAC
Th1= TIC 246M
D1,D2= 1N4004
L1=   40µH
K1,K2= 3-terminal connectors

Controlling the speed of a motor is quite simple. The voltage driving the motor is varied by a thyristor by changing the phase of its triggering pulse. A difference of few volts between the gate voltage of the thyristor and the cathode voltage (the motor voltage) determines the moment when the thyristor fires.

The rotation of the motor induces a counter-EMF with a force that is proportional to the motor's speed but opposite to the polarity of the supply voltage. When the motor rotates freely without load, the counter-EMF is equal to the supply voltage. In this case, the thyristor blocks the whole half wave swing long enough that the energy delivered to the motor is just sufficient to overcome the mechanical and electrical resistance. Once the motor is loaded, the speed will go down, thereby reducing the counter-EMF. At this moment, the thyristor conducts longer and delivers more energy so that the desired speed is regained and maintained.

Perfect speed compensation is only possible if the selected speed is far below the nominal value. As the selected speed comes closer to the maximum speed of the motor, the compensation becomes more difficult.

In constructing the circuit keep in mind that circuit works with hazardous voltages. A plastic box must be used as housing. The potentiometer must have a plastic handle and the knob must also be plastic. The thyristor must be heatsinked. Capacitor C1 must be able to handle 400 volts.

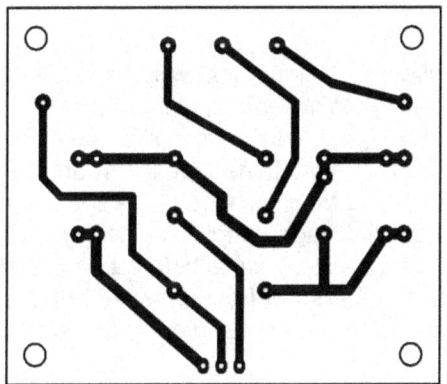

Figure 19.0 Printed Circuit Layout

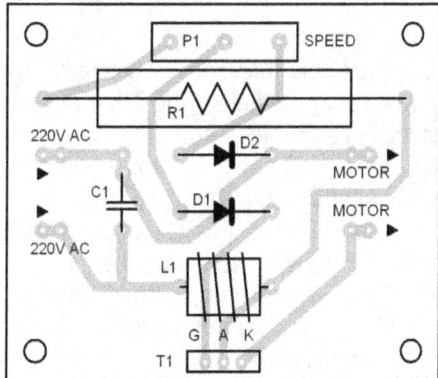

Figure 19.1 Printed Circuit Layout

Caution! Danger of electrocution!
Extreme shock hazard! You are working
with a line voltage of 220 Volts AC.

20   INTERCOM

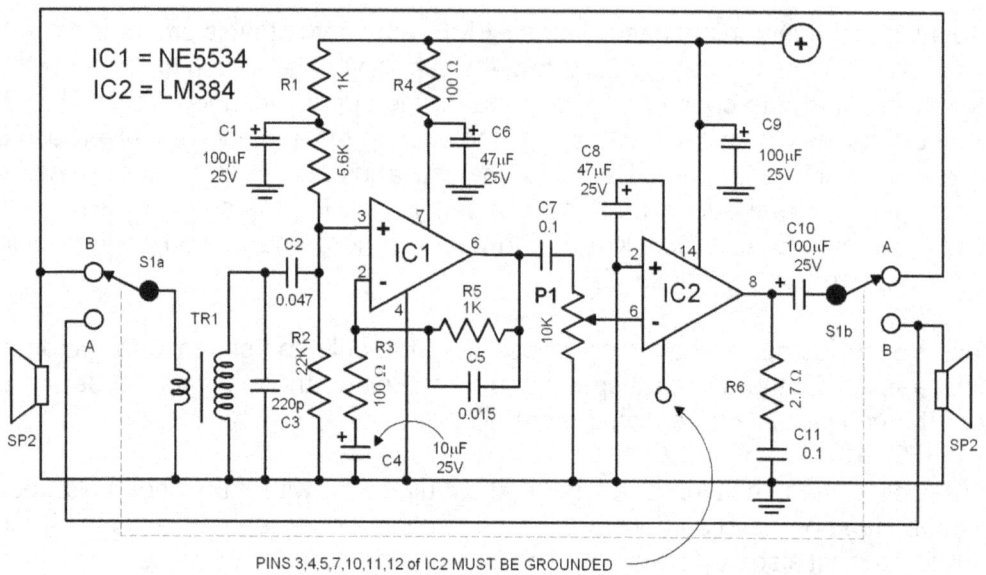

IC1 = NE5534
IC2 = LM384

PINS 3,4,5,7,10,11,12 of IC2 MUST BE GROUNDED

Diagram 20.0 Intercom

If you compare this intercom to modern FM and wireless designs, you will certainly say it is outmoded. Well..not quite right. The circuit fulfills its task - to provide a reliable communication line - and is very simple to construct. And that's what really counts. The circuit is made up of an amplifier, two switches, and two loudspeakers. If additional stations (speakers) are wished, additional switches are just incorporated into the circuit. IC LM382 is used as the final amplifier. It delivers nearly 2 watts of audio power by 15 volts supply voltage. Amplification is preset to 50. C9 serves as supply bypass. Another IC is used to boost the input signal before it is passed into the LM384. IC2 is an op-amp with an amplification factor of 11. The frequency bandwidth is 160 Hz to 10 kHz. These values were chosen since the device is to be used only as intercom and not as HIFI amplifier.

To achieve the best performance, use speakers that can also function as microphones. The intercom must be constructed inside a box so that the microphone function is optimum. Since the impedance of most speakers are quite low, an impedance converter is needed to maintain a good audio quality. This is done by the transformer TR1. TR1 is a normal 220/6V step-down transformer. The 6V winding of the transformer is connected to the speaker, and raises its impedance to about 10 Kohms. To reduce signal loss in the transformer, a 4.5 Watt transformer is used.

In constructing the intercom, house the power supply unit separately from the main circuit to avoid interference. C1 suppresses HF interference. The current consumption is about 210 mA with 1.8 watt output.

Page 44

21 TEMPERATURE ALARM

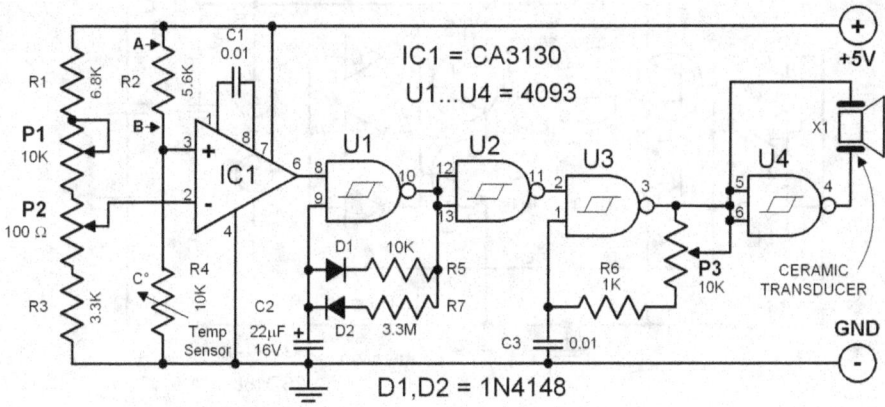

Diagram 21.0 Temperature Alarm

This alarm circuit continuously monitors the room temperature, and beeps when the temperature drops below 20 degrees centigrade. This capability to constantly monitor the room temperature can help lower your air-conditioning costs by reminding you when to turn off your airconditioner. An optical display like a thermometer has a limited effectivity since it can be easily overlooked. A tone signal on the other hand is not easy to ignore.

Let us look at the circuit. R4 is the temperature sensor KTY10. Other temperature dependent resistor types can also be used in its place. This sensor is connected to a resistance-bridge circuit and supplied with 5 volts. The IC1 CA3130 works as a bridge amplifier. As long as the temperature is higher than the value preset by P1 and P2, the output of IC1 is 0 volt. Once the temperature drops below the preset value, the plus input of IC1 receives more current, and its output level swings to 5 volts, thus switching the oscillator gate U1 on. This gate creates a pulse of around 0.2 seconds. This pulse in turn switches on the oscillator gate U3. The tone signal is generated by a piezo-electric crystal which is driven by gate U4.

The frequency of the tone signal is around 5 kHz. The current consumption is 2 mA which allows the circuit to be powered by a simple power supply. The sensor must be installed outside the housing so that is will not be falsely triggered by the heat of the transformer.

To align the circuit, points A and B must be shorted temporarily so that the alarm beeps. Potentiometer P3 controls the volume of the beep. The chosen room temperature can be roughly tuned by P1. P2 is used as a fine tuner. In normal operation, remove the short between points A and B.

22 ELECTRONIC DOG WHISTLE

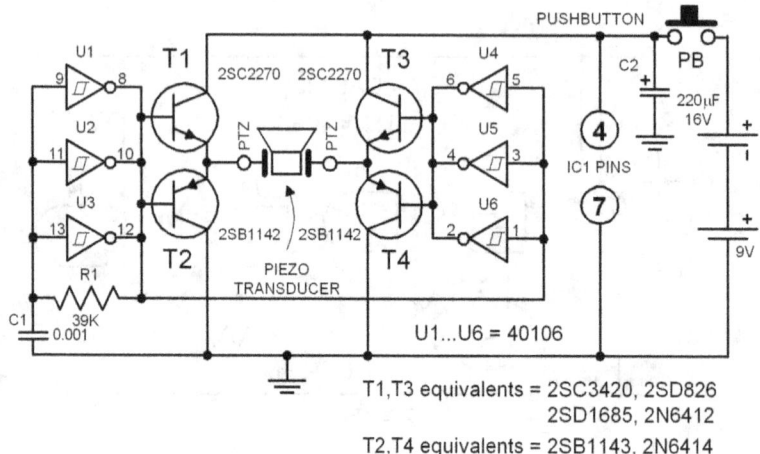

Diagram 22.0 Electronic Dog Whistle

It is widely known that dogs can hear sounds with frequencies higher than what humans can hear. Sound frequencies higher than 20 kHz can be hardly detected by humans but animals are very sensitive to it. Basing on this fact, an electronic dog whistle can be constructed and used to call dogs back from long distances.

The circuit featured here uses a piezo-ceramic element to generate the necesssary high audio frequency. The oscillator generates a squarewave signal which consumes less current than a sinewave. The oscillation is produced by gates U1...U3, R1 and C1. These components work as an stable multivibrator in the circuit. Since the piezo-ceramic introduces a capacitive load into the circuit, it generates high frequency spikes during voltage level changes. Transistors T1,T2 and T3,T4 serve as signal amplifier. Gates U4...U6 invert the signal from gates U1...U3 and create a "bridge". The circuit is powered by a single 9volt battery and generates a frequency of about 21 kHz.

This dog whistle generates an output voltage of about 15 volts p-p. It sends out a sound signal of around 101 dB! One word of caution in using the whistle. Small babies are found to be sensitive to this sound although mature people don't hear it. So, never use the whistle near babies.

 *Never use the whistle near babies!*

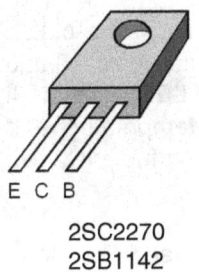

E C B

2SC2270
2SB1142

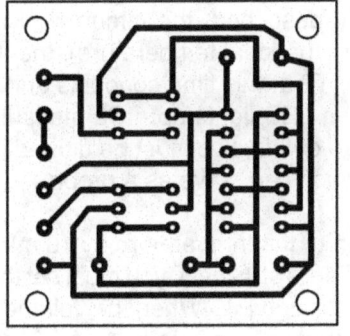

Figure 22.0 Printed Circuit Layout
for Electronic Dog Whistle

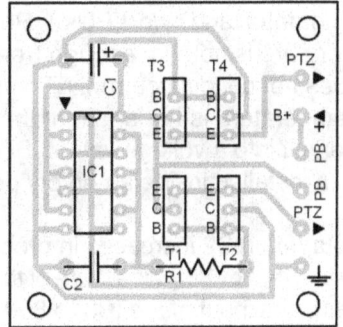

Figure 22.1 Parts Placement

23 COFFEE THERMOMETER

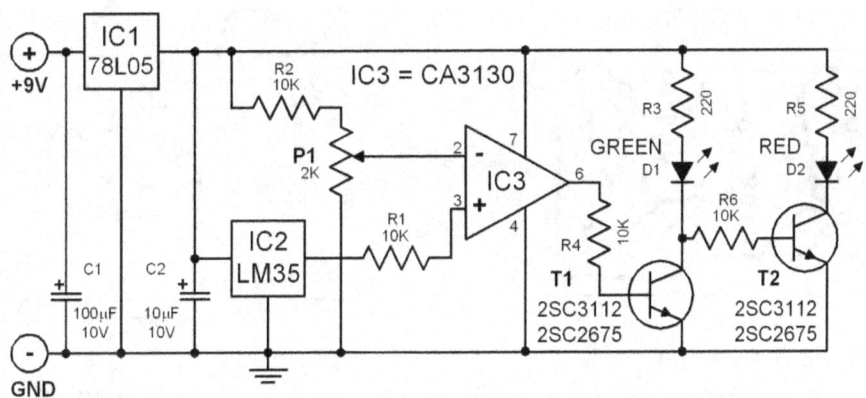

Diagram 23.0 Coffee Thermometer

Why measure the temperature of the coffee? Why not just drink it? Well, coffee taste depends on two things. First, how strong the coffee is and second, how hot it is. The first criterium is difficult to define since each person has his own "taste". Some wants the coffee to be very dark as if half of it is coffee powder. Others want to still see the bottom of the cup. The second criterium is quite easy since most of the coffee drinkers prefer to have it hot, that is, around 80 degrees centigrade. Since not everybody has a built-in thermometer in his tongue, the circuit featured here is very useful to save you from burning your tongue.

Electronic Circuits 1.2

The coffee thermometer is composed of a voltage regulator, a temperature to voltage converter, a comparator and two LEDs. Roughly described, the circuit works this way: If the coffee temperature is not hot enough (read: not the ideal temperature), the IC output is logical 0 and T1 does not conduct current. Transistor T2 at this time conducts and the red LED lights up. If the temperature is hot enough, the green LED lights up. IC2 measures the temperature and converts it into a voltage value. Because of this, IC2 must be dipped into the coffee. IC2 can also be installed inside an empty pen which can serve as a probe.

The output of IC2 increases in proportion to the temperature by 10 mV per degree. Therefore, if you feel that by 80°C the taste of the coffee is at its best, you must set the reference temperature at the minus input of the IC3 to 800 mV through P1. When the voltage level at the plus input of IC3 reaches 800 mV, the comparator output swings to logic 1 and T1 conducts current. At this time T2 turns off and the red LED is off. Otherwise, the green LED lights up showing that the coffee tastes good. There are two PCB layouts for this circuit. Choose any of the two.

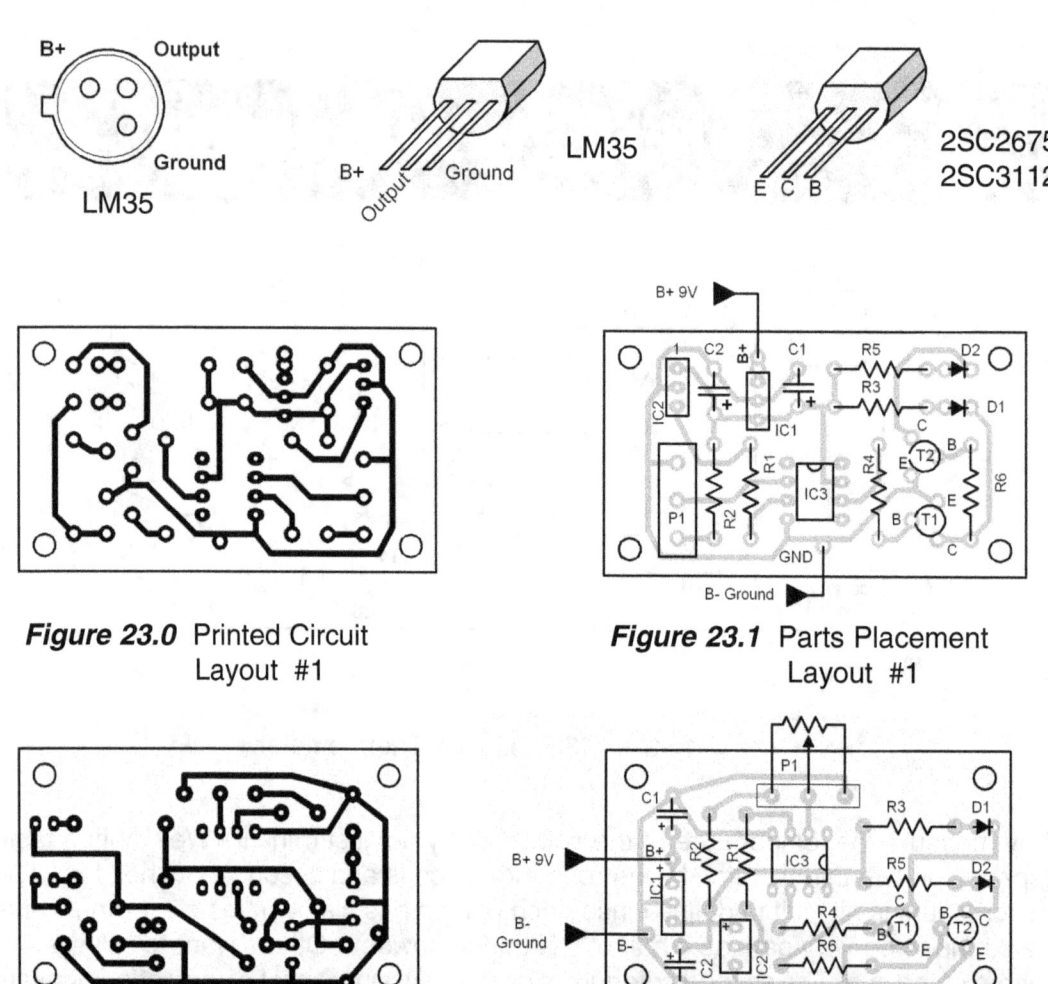

Figure 23.0 Printed Circuit Layout #1

Figure 23.1 Parts Placement Layout #1

Figure 23.2 Printed Circuit Layout #2

Figure 23.3 Parts Placement Layout #2

24 TELEPHONE LAMP CONTROL

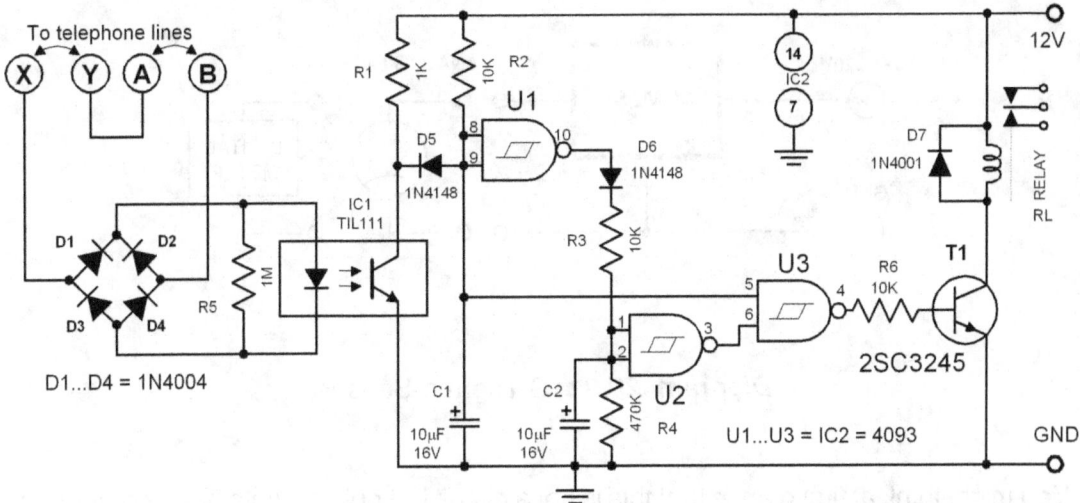

Diagram 24.0 Telephone Lamp Control

This circuit was originally developed for people who are hard-of-hearing. The circuit turns on a lamp when the telephone rings. Using this technique, the acoustic signal is practically converted to an optical signal. The design of the circuit featured here is very simple that it can probably be used for applications other than the one described here. The relay is held on for a while after the telephone handset is put down or after the ring has stopped.

A photo-coupler device is used in this design since the circuit is directly connected to the telephone line where voltage levels are high. During a ring or when the telephone is used, current flows through the rectifier D1...D4 and turns on the LED. The photo transistor conducts, C1 discharges and U3 logic reverses causing T1 to conduct current and the relay to close. After the handset is put down or the ring has stopped, C2 discharges slowly through R4 and the relay remains closed for a few seconds more.The current consumption is 10mA.

 *The coil voltage of the relay must be equal to the supply voltage.*

25 °C THERMO-SENSOR

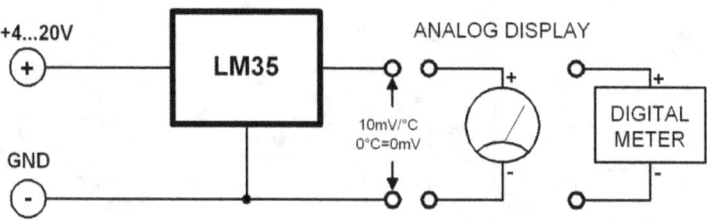

Diagram 25.0 °C Thermo-Sensor

You might think at first glance that this is not a circuit but only a single basic component. Well, it looks like a single component but it is actually a complete electronic thermometer. You only need to connect it to either an analog or digital display to see the actual temperature. The LM35 is a temperature sensor IC. Its voltage output is directly propertional to the temperature. The proportionality factor is 10 mV/°C, that means, when the temperature is 0°C, the output of the IC is also 0 volt. When the temperature is 18.5°C, the output of the IC is 185 mV. This is the big advantage of the IC in comparison to other sensors. You don't need to use external amplifiers or converters anymore. It consumes only 60 µA current and can be powered from 4 to 20 volts.

The output impedance is low which enables it to be directly connected to a coil-operated analog meter. In a practical application however, you must ensure that the internal resistance of the "load" connected to the IC is not less than 5 K-ohms to avoid overheating the IC. The measurement error at 25°C is typically 0.4°C. The IC can be directly connected to the analog input of a personal computer or a digital multimeter.

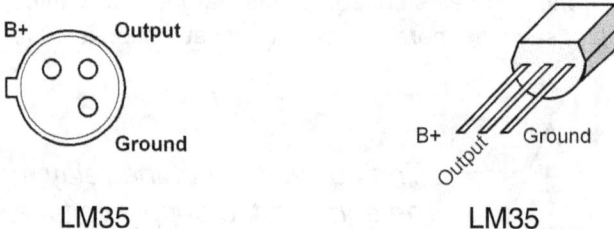

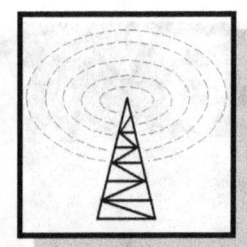

RADIO FREQUENCY

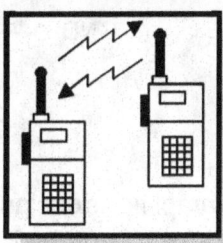

52 Cable TV Amplifier
53 2-Transistor Radio
54 Sensitive FM Tuner
55 CW Filter
56 Pin Diode Switch
58 Field Strength Meter
59 Mini AM Radio
60 Ocean Emergency Radio
61 FM Antenna Amplifier
63 Morse Code Filter
64 Shortwave Preselector
65 UHF Antenna Amplifier

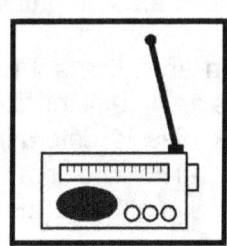

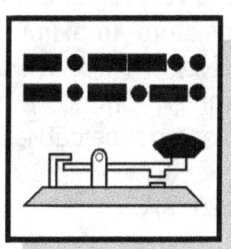

26 CABLE TV AMPLIFIER

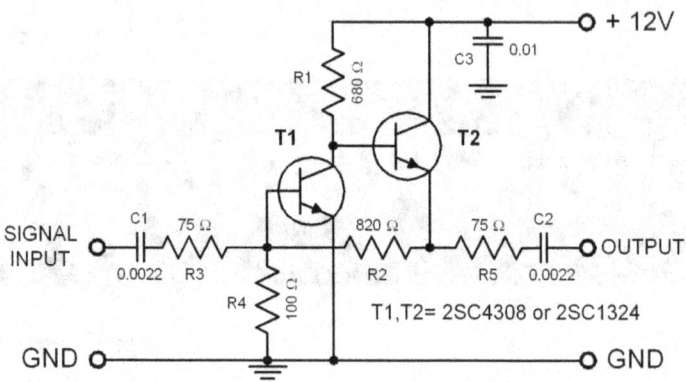

Diagram 26.0 Cable TV Amplifier

This RF-amplifier is designed to be quickly installed between two cables. A characteristic which is very important in repair and measurement jobs. Both input and output impedances are compatible with 75 ohms cables.

The main amplifier is T1. Transistor T2 works as an emitter follower. The feedback bias is determined by R3 and R4. The circuit amplifies the incoming signal by 22 dB.

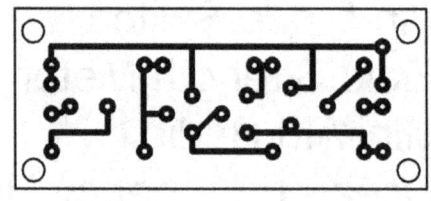

Figure 26.0 Printed Circuit Layout

Due to the very high frequency limit of the transistors (maximum= approx. 2,000 MHz), the circuit works well up to 150 mHz. It must be housed in a metal case and the cables must be 75 ohm type. The amplifier consumes around 20 mA.

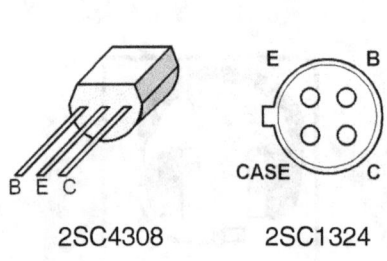

2SC4308   2SC1324

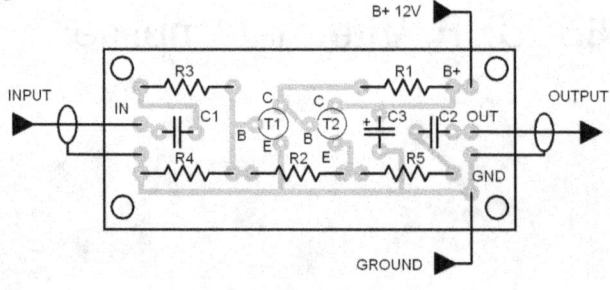

Figure 26.1 Parts Placement Layout

27 2-TRANSISTOR RADIO

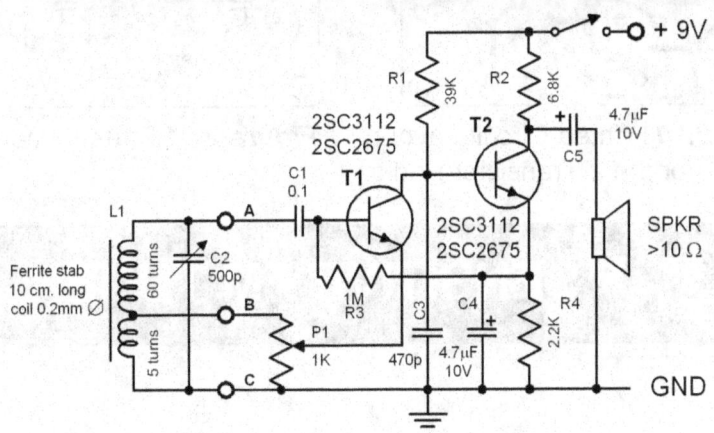

Diagram 27.0 2-Transistor Radio

This circuit is also called "mini-radio". It uses only two transistors and few passive components which makes it very easy to be constructed. Although the circuit is very simple, it functions very well without external antenna or ground connection. The transistor T1 works as a feedback regulated HF-amplifier and functions as demodulator at the same time. The sensitivity of the receiver is dependent on the amount of feedback and can be adjusted by P1.

Parts List

Resistors:

R1= 1 M-ohm
R2= 39K
R3= 6.8 ohm
R4= 2.2 ohm
P1= 1K potentiometer

Capacitors:

C1= 500 pF - variable capacitor
C2= 0.1mF
C3= 470p
C4,C5= 4.7μF/10V Electroytic

Transistors:

T1,T2= 2SC3112(2SC2675)

The demodulated signal comes out from the collector of T1. The signal is then filtered by C3 so that only the audio signal will be amplified by T2. The amplified signal is then delivered to a high impedance "earphone". The coil is 65 turns AM antenna wire around a 10 cm long x 10mm diameter ferrite rod. The tap is at the fifth turn of the coil counting from its ground end. The coil must be installed as close as possible to the PCB.

The sensitivity of the receiver can be greatly improved by attaching an external antenna into it. The external antenna must be coupled to the hot end of the coil through a 4.7 picofarad capacitor. The receiver can be powered by a 9 volt battery . It consumes only 1mA.

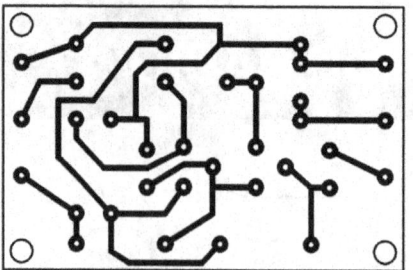

Figure 27.0 Printed Circuit Layout for the 2-Transistor Radio

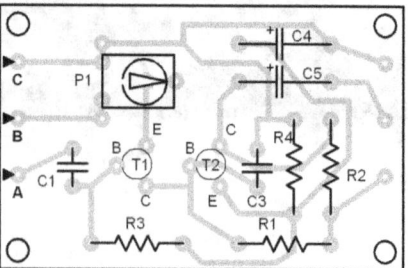

Figure 27.1 Parts Placement Layout

28 SENSITIVE FM TUNER

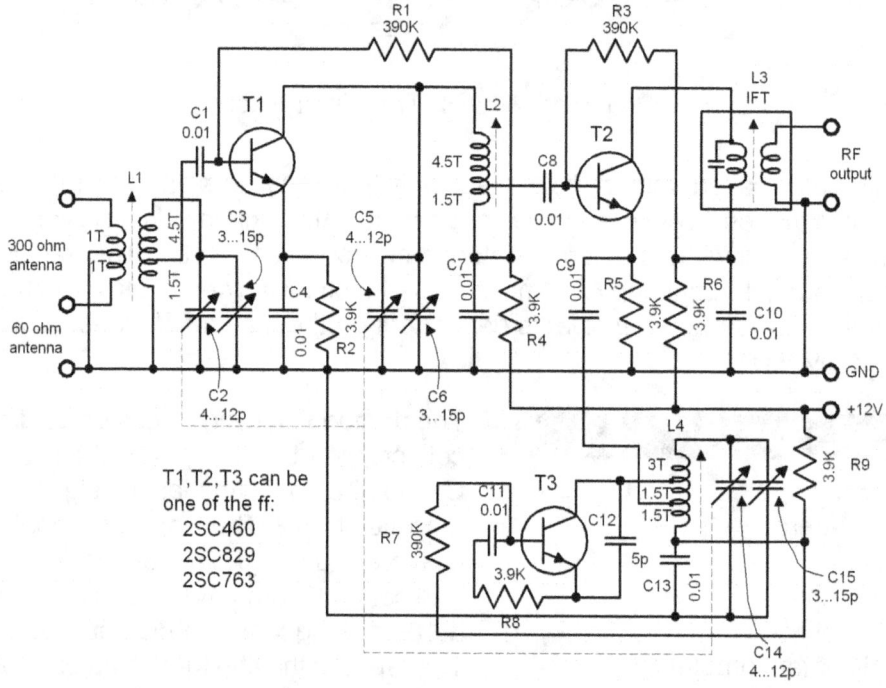

Diagram 28.0 Sensitive FM Tuner

This is the right circuit for hobbyists who want to build their own tuners instead of buying a "plug-n'-play" finished product. The tuner circuit is designed with only 3 transistors. The amplification is around 40 dB. The first transistor works as an RF amplifier. The second transistor is the mixer. The incoming signal is introduced into the base and the oscillator signal is coupled to the emitter. The third transistor is the oscillator. The coils are wind around 6mm coil formers with a ferrite core. The best material for the coils is a silver coated copper wire since it is very easy to be tapped. The distance between the windings must be 0.8 mm.

29 CW FILTER

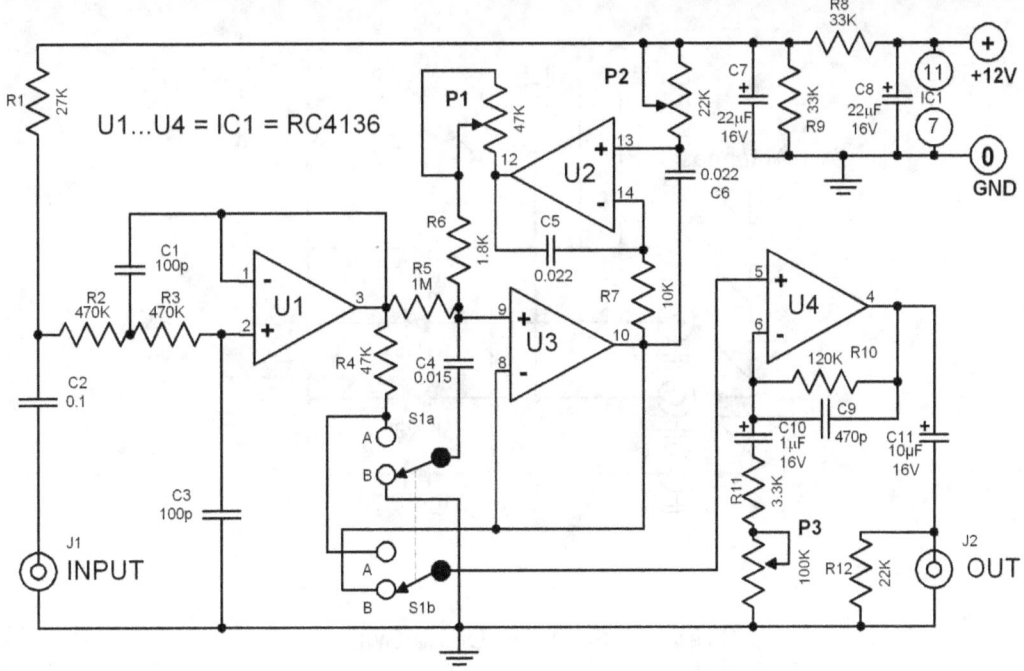

Diagram 29.0 CW Filter

Cheap shortwave receivers have a disadvantage of -along with low sensitivity- having poor selectivity that results to an audio signal distorted by interference. The interference is usually caused by the mixing of the carrier frequency (which was not completely suppressed by the bandfilter) with the signal of the CW modulator. This kind of interference can be filtered out through the use of additional Band or Notch filter in the audio range.

The filter circuit featured here can be used either as notch filter or as bandpass filter. Switch S1 selects the function of the filter. When S1 is switched to position A, the circuit functions as a notch filter. In this case a small portion of the band is not allowed to pass through the filter. This portion is normally the interfering signal and is selected out by adjusting P2.

When S1 is at B position, the circuit functions as a bandpass filter allowing a narrow portion of the band to pass through it. The center frequency of the band can be adjusted by P1. The output of the filter can be directly connected to the speaker or tape input of the short wave receiver being used. A headphone with an impedance of more or less 600 ohms can also be directly connected to it.

30   PIN DIODE SWITCH

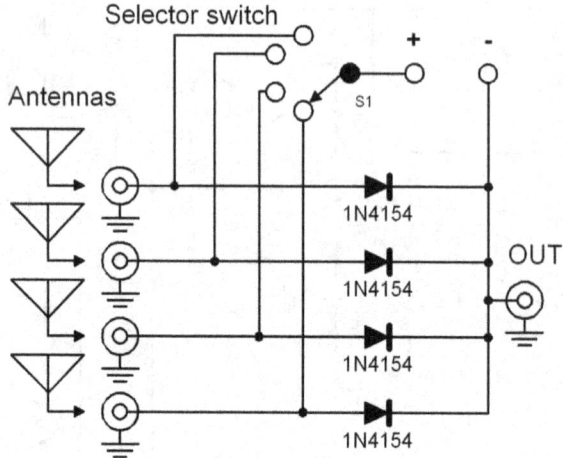

Diagram 30.0 Pin Diode Switch

A mechanical switch does not cause too much signal loss in low frequency area. In VHF and UHF area however, the scene changes drastically.

The ideal antenna switch for VHF and UHF work is the PIN diode antenna switch. PIN diodes are special high frequency switching diodes with very low internal capacitance. Furthermore, the internal resistance of a PIN diode can be remotely varied by a DC control voltage from 1 up to 10,000 ohms. The resistance of a PIN diode changes linearly in relation to the current flowing through it. The higher the current, the lower the internal resistance. This characteristic opens a wide area of possible applications. If the control current is linearly controlled, the PIN diode can be used to weaken or totally shut off RF signals and to amplitude modulate a carrier signal. If the control current is periodically switched, the PIN diode can be used for pulse and phase modulation of RF signals.

In the featured circuit, the PIN diodes are used to shut on or off several antennas. The circuit is composed of a current source, four PIN diodes and a rotary switch with four positions. The circuit diagram itself explains almost everything. The current source is a 12 volt circuit composed of a transformer, bridge rectifier, and a regulator IC. In order to instantly know which antenna is switched on, 4 LEDs are incorporated into the circuit.

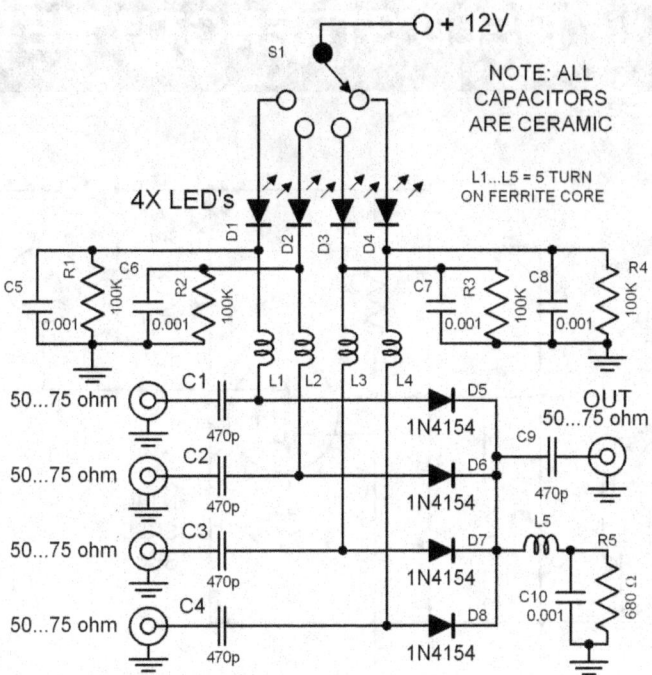

Diagram 30.1 Pin Diode Switch

First, the control current flows through whichever LED is switched on, then through the coil and the PIN diode, then finally through coil L5 and resistor R5. Resistor R5 determines the amount of current flowing through the PIN diode. With a value of 680 ohms, the current is around 15 mA. It is enough to control the PIN diode and at the same time power the LED.

Capacitors C1...C4 and C9 are DC blocking capacitors. The choke coils L1...L5 prevents the RF signal from flowing to ground through the current source circuit. Capacitors C5...C8 are bypass capacitors. Resistors R1...R5 are pull-down resistors so that the disconnected LEDs will stay off.

In constructing the circuit, all wirings must be as short as possible to avoid picking up interference. Coils L1...L5 are 2 turns magnet wire around a ferrite core for VHF range. For UHF range, the coil must be 5 turns. If you want to use ready-made choke coils, use 1 µH for UHF range and 5µH for VHF range.

The switch can be used for antennas with impedances of from 50 ohms up to 75 ohms. The opened diodes block the "disconnected antennas" by about 30 dB. Signal losses at the closed diode is negligible. The signal to noise ratio of the receiver however deteriorates somewhat up to 1 dB.

31   FIELD STRENGTH METER

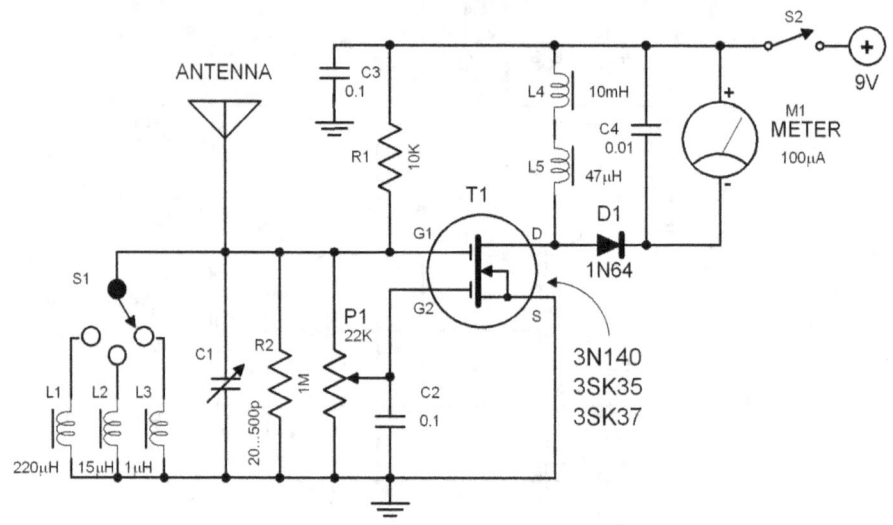

Diagram 31.0   Field Strength Meter

A sensitive and reliable field strength meter is an invaluable instrument in amateur radio as well as in radio-controlled model area. A field strenght meter is used to align an antenna to get the best possible gain, to choose the "right" commercial antenna, to determine the transmitting range of radio controllers, etc.

Designing the field strength meter to be sensitive is due to three reasons. First, there must be as many wavelengths as possible between the meter and the transmitter. Second, the measurement can be done without using a stronger carrier signal. Third, most of the radio controllers are of low power type. Considering these factors, a dual gate MOSFET is used as an RF amplifier in the circuit. The amplification can be adjusted by P1.

Switch S1 selects three frequency ranges:

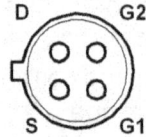

3SK35
3SK37
3N140

L1= 480 kHz to 2.4 mHz
L2= 2.4 mHz to 12 mHz
L3= 12 mHz to 40 mHz

The antenna is made of 30 cm long metal rod.

32   MINI AM RADIO

This AM receiver circuit is based on the IC ZN415 introduced by Ferranti. This IC is an improved version of the early ZN414. The IC is a complete AM receiver and does not need IF transformers and alignment. Let us look inside the IC. Integrated in the IC is the early ZN414 and an audio amplifier circuit.

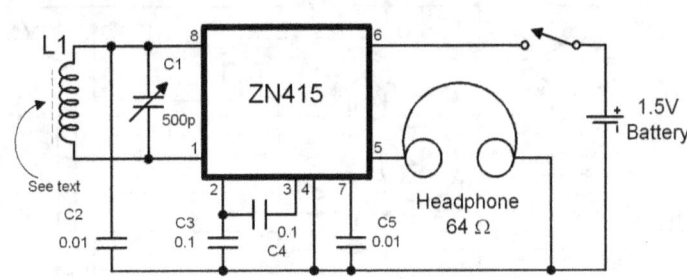

Diagram 32.0 Mini AM Radio

The tuner stage covers the frequency area from 150 kHz up to 3 mHz. It is designed to receive MW and LW signals.The audio output is around 1...1.5 mW at 64 ohmn load. The IC comes in an 8 pin DIL package. Originally, a headphone is enough to enjoy the functionality of the receiver but is sounds better by adding an external audio amplifier that can drive an 8 ohm speaker. This circuit however needs 9 volts supply in contrast to the IC which needs only 1.5 volts. The power output of the amplifier circuit is around 120 mW. The headphone can be left connected to the IC while the external amplifier works since it is connected to a separate pin of the IC.

If the circuit is to receive MW signals, the coil L1 must be 65 turns of antenna wire around a 600mm x 12mm x 3mm flat ferrite bar. If it is to receive LW signals, the coil L1 must be 300 turns of antenna wire around 150mm x 10mm ferrite rod. The coil for MW operation is wind in a single layer. The coil for LW operation can be wind on top of each other. To enable reception on both frequency bands, a 10pf capacitor must be connected in parallel to the LW winding. A switch must then be used to select the necessary coil.

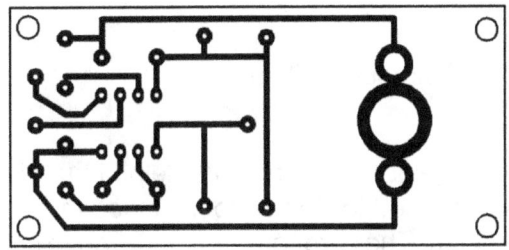

Figure 32.0 Printed Circuit Layout

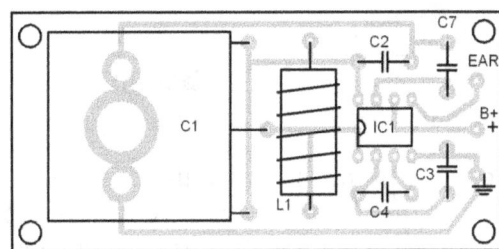

Figure 32.1 Parts Placement Layout

33 OCEAN EMERGENCY RADIO

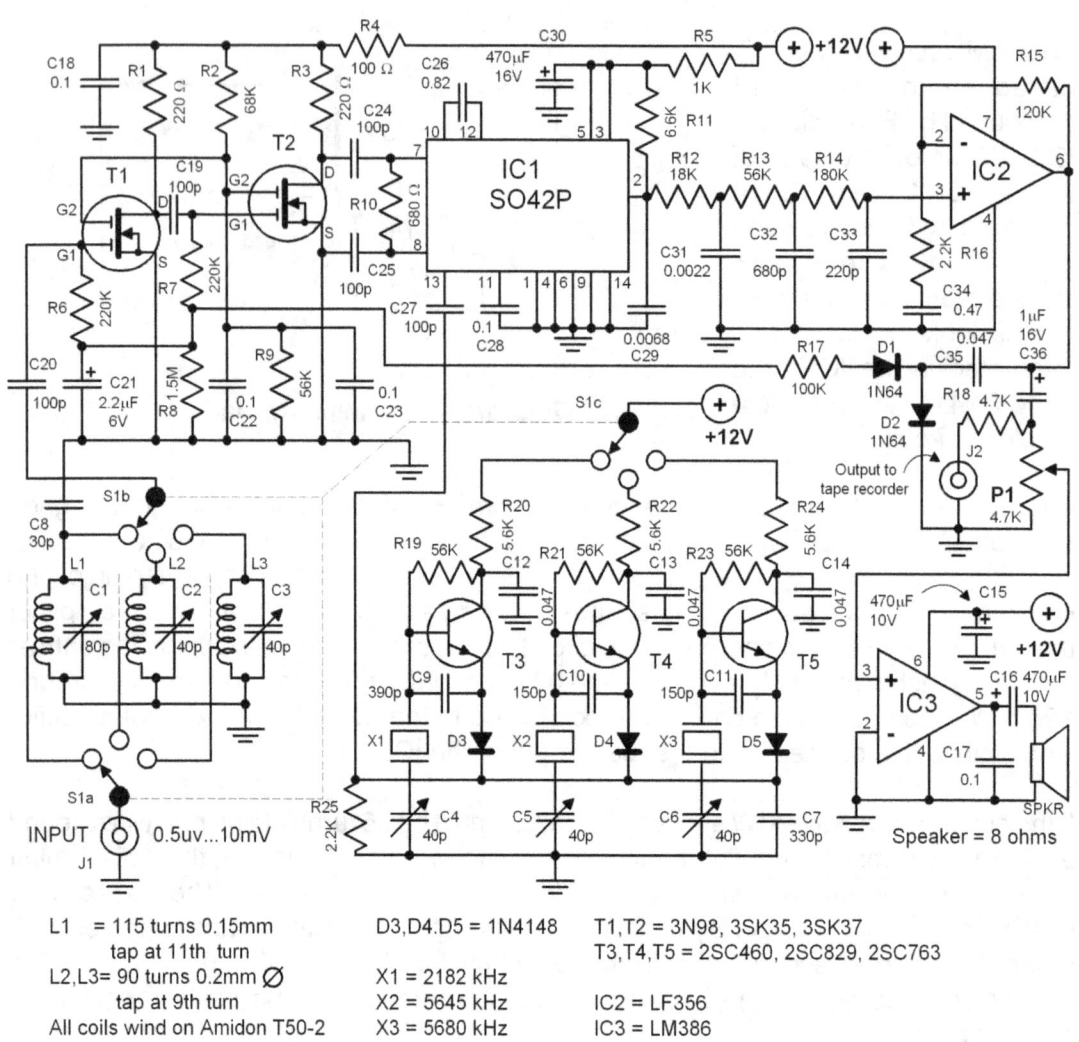

Diagram 33.0 Ocean Emergency Radio

Continuous monitoring of the ocean emergency frequency is not just helpful but also necessary. Single channel receivers are usually used for this purpose since they do not need to be tuned or aligned every now and then. They are "always ready" so to say.

The receiver circuit featured here uses the so-called "direct-conversion" method of receiving. It is a frequency converting receiver with an oscillator frequency which is the same with that of the received frequency. Single channel receivers normally employ this signal conversion technique because of its relative simplicity.

The received signal first enters the input circuit composed of two MOSFETS. The dual gate MOSFETS offer a very effective means of regulating the signal's amplitude. The IC works as a mixer where the received signal and the oscillator frequency are mixed. The detected audio signal comes out of pin 2, filtered by the succeeding LC circuit and then amplified by IC2. IC3 functions as a final amplifier to enable the audio signal to drive the loudspeaker.

As stated before, the oscillator frequency is the same with the received frequency. In this circuit the oscillator frequency is also the crystal frequency. If you want to add crystals or change the existing crystal with another frequency, the coils and trimmers at the input circuit must be redesigned. The coil's inductivity and the capacitance of the trimmers and capacitors are inversely proportional to the crystal frequency.

In constructing the circuit, keep the wirings as short as possible. The dashed lines in the diagram show where to put a metal shield. The receiver can function well up to 18 mHz. The alignment of the receiver is very easy. First, the capacitors C4,C5 and C6 must be tuned to the crystal frequency. Second, the capacitors C1, C2, and C3 must be tuned for maximum audio output.

34 FM ANTENNA AMPLIFIER

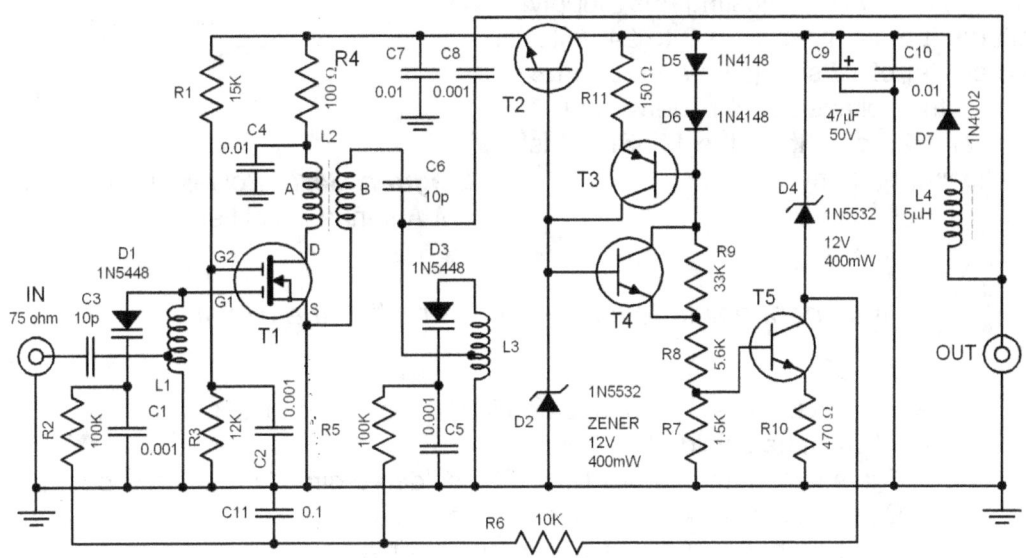

Diagram 34.0 FM Antenna Amplifier

T1 = 3N225, 3SK85, 3SK87	L1 = 9 turns 0.7mm ⌀ tap at 1st turn from the ground
T2 = 2SD781, 2SD1177, 2SD1684,	L2 = 613 turns (A & B) 0.5mm on Amidon T37-12
MJE243,MJE244	L3 = 9 turns 0.7 mm⌀ tap at 3rd turn from the ground
T3 = 2SA970, 2SA1136, 2SA1137	
T4,T5 = 2SC3622, 2SC3245, 2SC3248	

Table 34.0 Transistor and coil data

In some areas the signal reception of FM stations is too bad that in order to receive a satisfactory signal one must install an antenna amplifier.

This circuit is just the right thing to do it. It is designed in such a way that the supply current powering it flows through the coax cable. With this technique, an extra cable to power the amplifier is unnecessary. The RF signal and the DC current supplying the amplifier use the cable simultaneously. The RF signal is however prevented by LC filters from flowing into the power supply. The amplifier works with either 50 ohm or 75 ohm antenna. It has a gain of 25...30 dB. MOSFETS are used to avoid the problem of crossmodulation or intermodulation. The circuit is divided into two parts. The first part is the active amplifier circuit which is normally installed very near the antenna. The second part is the power supply circuit which supplies current to the amplifier through the coax cable.

After the circuit is constructed, the power supply should be connected to it. You must then test the DC voltage in the cable. This must be between 15.5 and 36.6 volts. Then test the voltage between T5 collector and ground while adjusting the supply voltage. It must move from 3 to 24 volts. The emitter of T2 should be approx. 11.4 volts. If the voltage at R4 is between 0.7 volts and 2 volts, then the MOSFET is functioning properly.

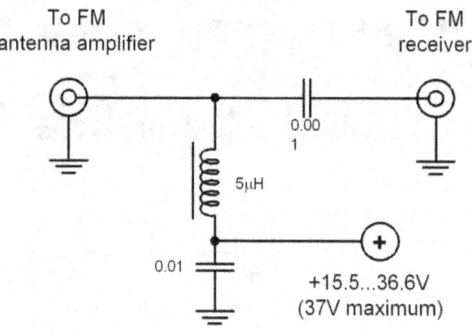

Diagram 34.1 Power Supply for the AM Antenna Amplifier

 *Important points to follow in constructing the circuit:*

- The circuit must be constructed in a double sided PCB.
- T1 must be shielded.
- The source terminal of T1 must be soldered directly into the copper plate.
- C4 terminals must be as short as possible.
- Coil terminals must be as short as possible.
- The antenna must be connected directly to L1.
- T2 must be heatsinked.

35 MORSE CODE FILTER

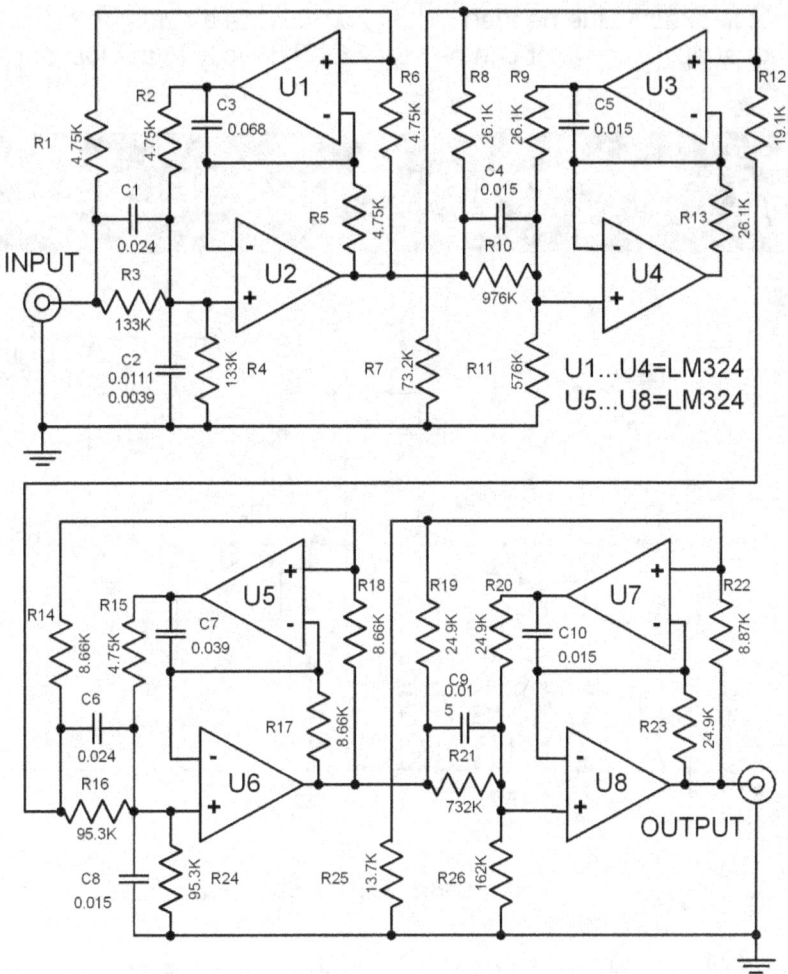

Diagram 35.0 Morse Code Filter

Morse code telegraphy is a very interesting technique of communication. Many radio amateurs are using this mode because morse telegraphy has longer range than voice modulation and because morse transmitters are simple and cheap to construct. Morse code communication is however sometimes plague with interference problems. The reception through heavy interference can be very bad especially when simple receivers are used.

Electronic Circuits 1.2

The filter circuit featured here is designed to help improve the quality of signal reception. Actually the filter circuit does not improve the ability of the receiver to receive but instead it filters out unwanted interfering signals from the audio to make the morse signal more readable. The narrow bandpass characteristic of the filter lets only the signals within 380 Hz to 500 Hz pass through. To achieve this accuracy, the resistor values must be exactly that what the circuit shows. Sometimes you need to parallel two components in order to get the exact value needed. The circuit needs a symmetrical +15 volts and -15 volts power supply. The filter can be connected directly to the loudspeaker output of the receiver.

36 SHORTWAVE PRESELECTOR

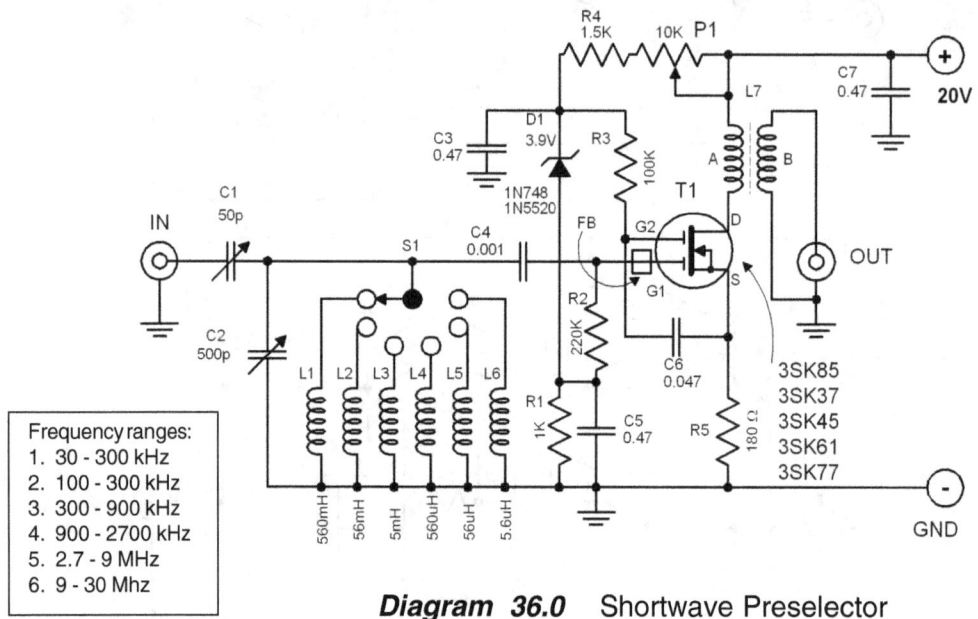

Frequency ranges:
1. 30 - 300 kHz
2. 100 - 300 kHz
3. 300 - 900 kHz
4. 900 - 2700 kHz
5. 2.7 - 9 MHz
6. 9 - 30 Mhz

Diagram 36.0 Shortwave Preselector

In comparison with preselectors of ordinary shortwave receivers, this circuit has an input voltage of up to 2.5V p-p and the output voltage is around 3V p-p with 50 ohm impedance. The low input capacitance of MOSFETS makes it possible to use feedback techniques through an unbypassed source resistor. This gives the preselector a very good dynamic range.

The overall amplification can be adjusted through C1. The maximum current consumption can be adjusted through P1 up to 12.7 mA. Wind the six input coils around ceramic coil formers with at least 10 mm diameter. The ferrite bead FB must be inserted directly into the gate terminal of T1 to prevent parasitic oscillation in VHF/UHF range. The output transformer is made of 20 turns for A and 4 turns for B magnet wire into a ferrite ring G.2-3/FT16.

37 UHF ANTENNA AMPLIFIER

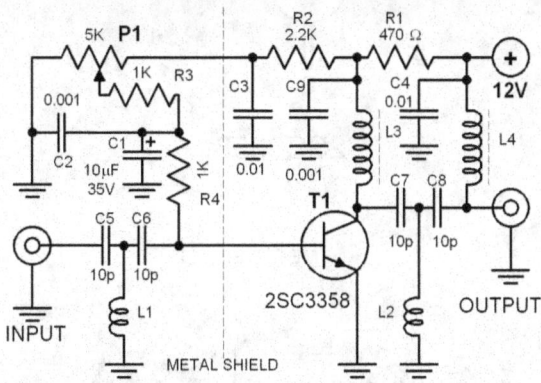

L1,L2 = Aircore 2 turns 0.5mm ∅3mm(coil diameter)
L3,L4 = 10 uH HF choke or 0.2mm ∅10 turns in ferrite core

Diagram 37.0 UHF Antenna Amplifier

This amplifier circuit is used to amplify TV signals in UHF range. It uses a low-noise transistor and gives 10 to 15 dB amplification in the frequency range from 400 MHz to 850 MHz.

As shown by the dashed lines in the diagram, the transistor must be shielded from the input components. In constructing this circuit, the wirings must be as short as possible. For best results, the cable must be directly soldered into the PCB. The circuit should be installed near the antenna and housed in a water proofed case.

The power supply for the circuit is fed through the cable by using a choke coil. To prevent the DC voltage from getting into the TV set, the coax cable must be coupled to the set through a small value capacitor.

To align the amplifier, just adjust P1 until the best reception is achieved. It means a collector current between 5 and 15 mA.

Parts List:
Resistors:
R1=   470Ω
R2=   2.2K
R3=   1K
R4=   1K
P1=   5K trimmer potentiometer
Capacitors:
C1=   10µF/35V electrolytic
C2,C9= 0.001µF/50V ceramic
C3,C4= 0.01µF/50V ceramic
C5,C6,C7,C8= 10pF/50V ceramic
Transistor:
T1,= 2SC3358
Coils = see text

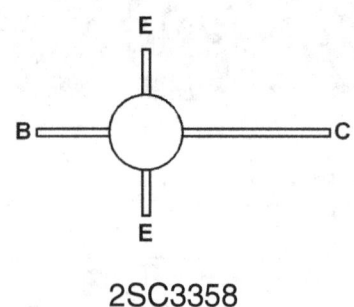

2SC3358

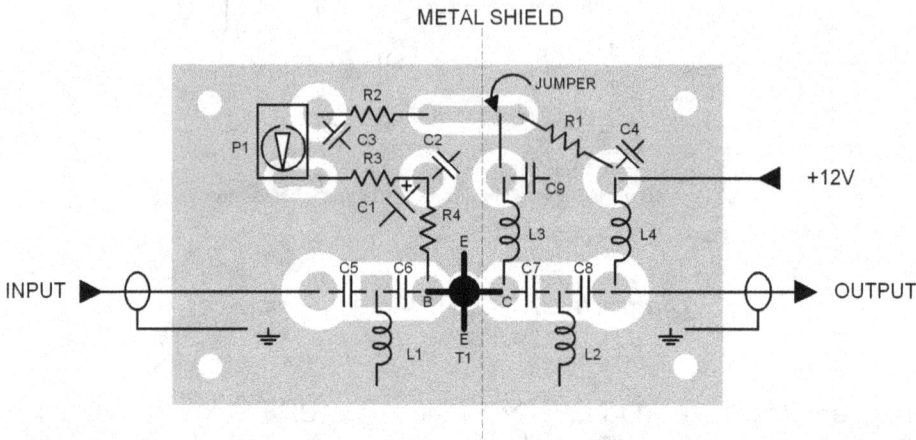

Figure 37.0 Parts Placement Layout for the UHF Antenna Amplifier

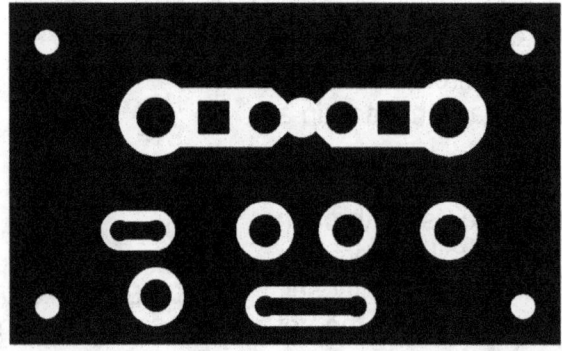

Figure 37.1 Printed Circuit Layout for the UHF Antenna Amplifier

HOBBY & SHOP

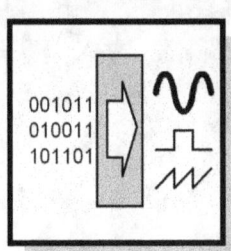

68 Nicad Battery Charger
70 Thyristor Signal Generator
71 Frequency Divider
72 Regulated Mic Preamp
73 Superzener Circuit
74 Audio Delay Line
75 Trigger Switch
76 Frequency Multiplier
78 Electronic Potentiometer
79 Electronic Selector Switch

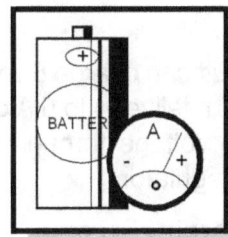

38 NICAD BATTERY CHARGER

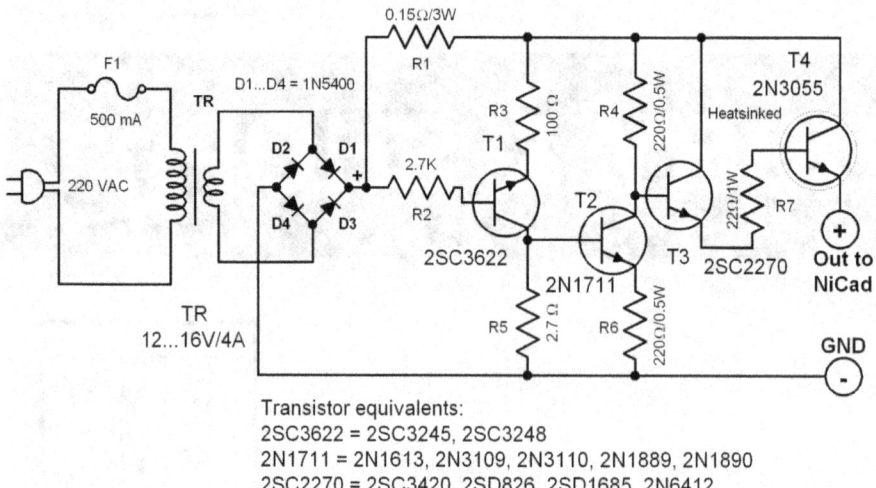

Diagram 38.0   NiCad Battery Charger

This circuit can charge 6 volts as well as 12 volts NiCad batteries. It uses a transformer which can deliver 4 to 5 A current between 12.6 and 16V. Ordinary chargers select the charging voltage through a switch but the circuit featured here uses an automatic current regulator.

The charging current is automatically regulated to 4.2 amperes. Once the charging current reaches 4A, the voltage drop at R1 will be 600 mV. In this case the transistor T1 will conduct and supplies current to T2.

Parts List:

Resistors:
R1=   0.15Ω/3W
R2=   2.7K
R3=   100Ω
R4,R6=   220Ω/5W
R5=   2.7Ω
R7=   22Ω/5W

Diodes:
D1,D2,D3,D4=   1N5400
TR = 220VAC/12...16V/4A
Transistors:
T1,T2,T3= See text
Fuse = 500 mA

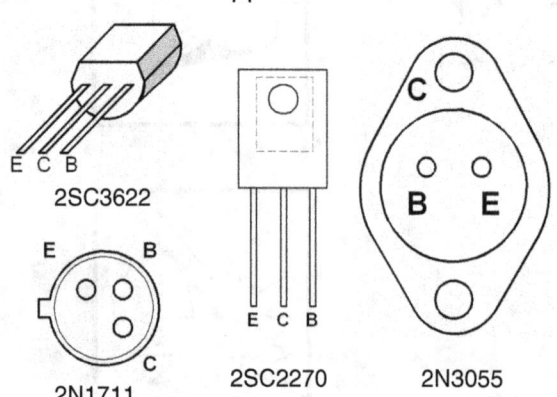

Transistor T2 in turn, shorts the base of T3 to ground thereby reducing or totally cutting off the base bias current of T4. The voltage difference between the collector voltage of T4 and the actual charging voltage of the Nicad battery is dissipated by T4. The power dissipation of T4 is therefore the product of the voltage difference and the charging current. In charging 6-volt batteries, the power dissipation can reach up to 40 watts.

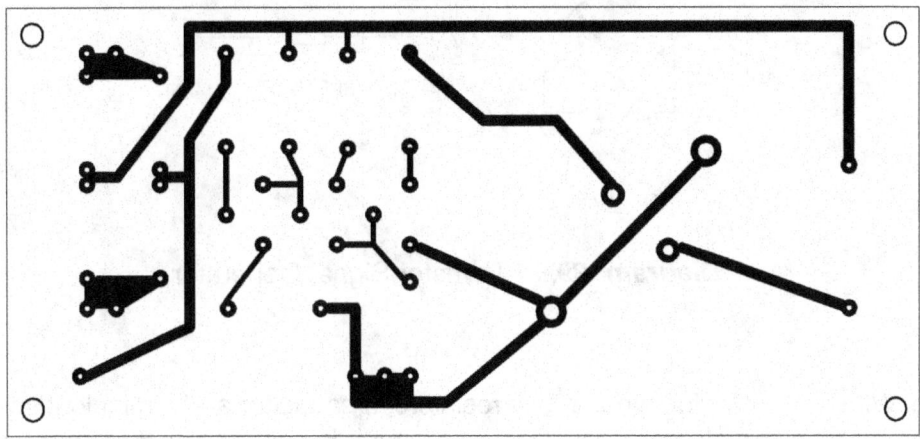

Figure 38.0 Printed Circuit Layout for the NiCad Battery Charger

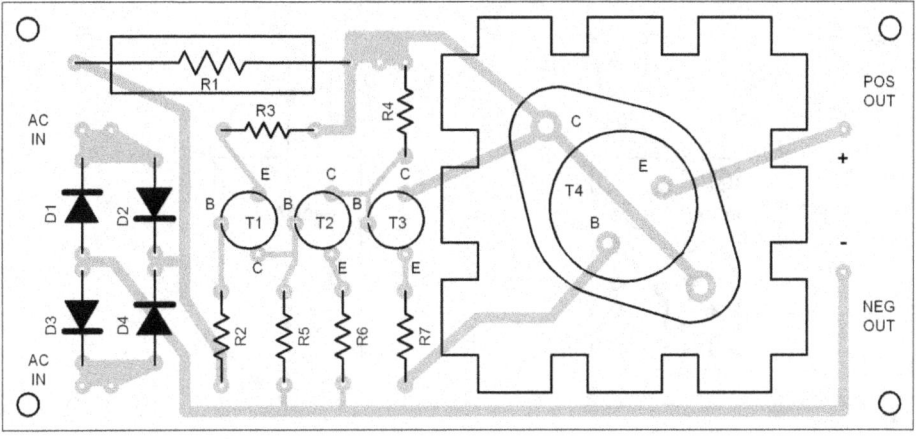

Figure 38.1 Parts Placement Layout for the NiCad Battery Charger

39 THYRISTOR SIGNAL GENERATOR

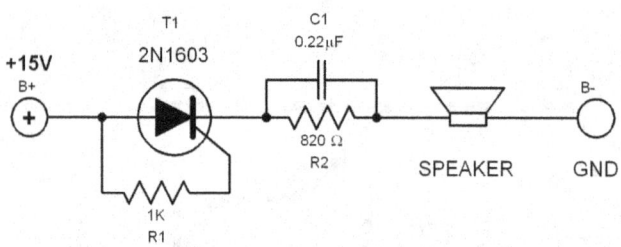

Diagram 39.0 Thyristor Signal Generator

Simple. With only a single thyristor, two resistors, a capacitor and a mini-loudspeaker, you can make a signal generator. Once the circuit is switched to a current source, current flows through R1 and switches on the thyristor. The thyristor conducts current and charges capacitor C. Since the charging current is inversely proportional to the charging of the capacitor, it slowly decreases until such a level that the thyristor switches off. Once the thyristor is off, capacitor C discharges through R2 and the whole process is repeated. The periodic switching on and off of the thyristor generates the oscillating signal. The frequency of the oscillation is determined by the C/R2 combination.

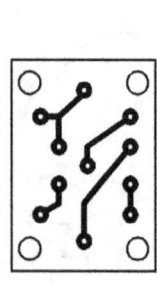

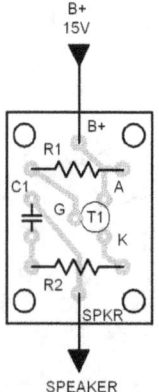

Figure 39.0
Printed Circuit Layout

Figure 39.1
Parts Placement Layout

40 FREQUENCY DIVIDER

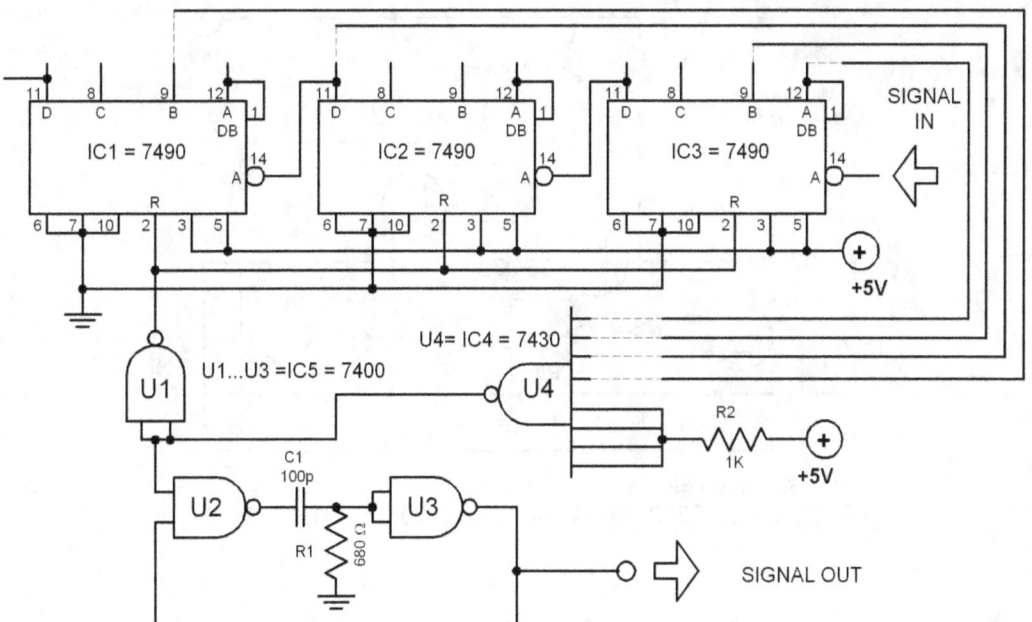

Diagram 40.0 Frequency Divider (Programmable)

This circuit divides the frequency of a TTL level input signal. The division factor can be programmed at will from 1 up to 999. It is composed of three decimal counters, one 8-input NAND gate and a pulse former (U2,U3). The incoming signal is fed to pin 14 of IC1. In programming the division factor, the necessary BCD output of the three counters must be connected to the inputs of the NAND gate. The unused inputs must be connected to the +5V line through a 1K pull-up resistor. The output of the 8-input NAND gate is inverted by U1 and connected to the reset pins of the counters.

The connection of the circuit in the diagram shows the sample division factor of 283. Once the counters reach the programmed number, all inputs of IC4 will be logic "1". IC4 then outputs a logic "0" which is inverted to logic "1" by U1. This logic "1" then resets all the counters and the cycle is repeated.

Gates U2 and U3 form a monostable multivibrator which sends out a short pulse everytime the counter resets. The output signal comes out from U3.

41 REGULATED MIC PREAMP

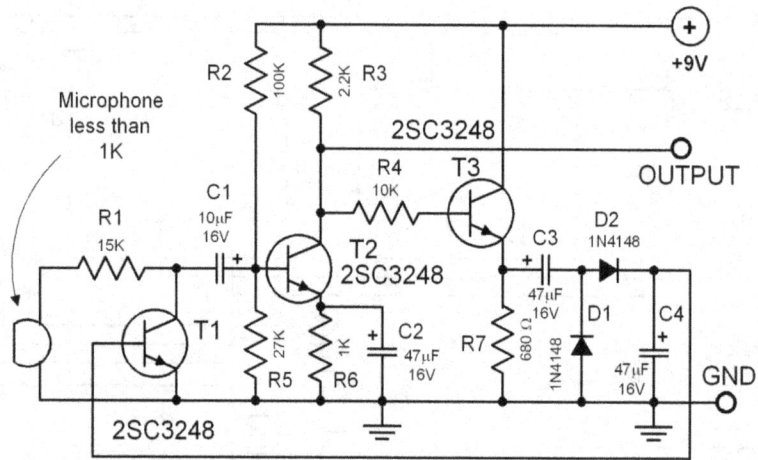

Transistor equivalents:
2SC3248 = 2SC3622, 2SC3245, 2SC2459, 2SC3112, 2SC2675

Diagram 41.0 Regulated Mic Preamp

This microphone preamp boosts-up weak audio signals before they are introduced into another circuit for further processing. This circuit is designed to be used particularly in modulating radio transmitters.

The audio signal picked up by the microphone is first amplified by T2 and passed on to T3 which functions as emitter follower. The signal is then rectified by diodes D2-D1 and smoothened by capacitor C4. This signal is then used as feedback to control T1. When the audio signal is very strong, T1 conducts more current thereby reducing the signal that reaches T2. The maximum input voltage is about 1 Vp-p. The microphone can also be replaced by a loudspeaker.

Parts List:

Resistors:
R1= 15K
R2= 100K
R3= 22K
R4= 10K
R5= 27K
R6= 1K
R7= 680Ω

Capacitors:
C1= 10μF/16V electrolytic
C2,C3,C4= 47μF/16V electrolytic

Transistors:
T1,T2,T3= 2SC3248

42 SUPERZENER CIRCUIT

This circuit is originally designed to be used in a battery powered device with a minimal current consumption. The change in the output voltage of the IC is only 1 mV although the power supply voltage is increased from 1 to 30 volts. The voltage drop at the zener diode is extremely constant when a constant current of around 1 mA flows through it. The current is determined by R1. The combination R1/D1 forms a kind of constant current source. The output voltage of the IC can be mathematically computed using this formula:

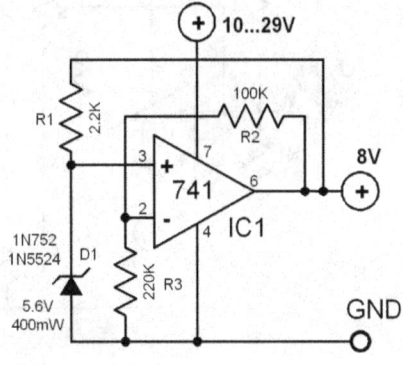

Diagram 42.0 Superzener Circuit

$$V_{out} = \frac{R2+R3}{R3} (V_{zener})$$

as long as the power supply voltage is 2 volts higher than the output voltage. The IC can supply a current of about 15 mA. This circuit does not compensate for temperature changes in the zener diode. If temperature is a critical factor in the application, use a zener diode with the lowest temperature coefficient.

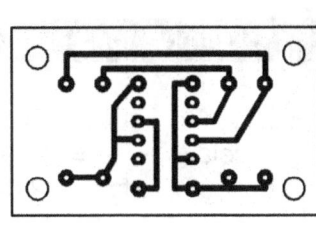

Figure 42.0 Printed Circuit Layout

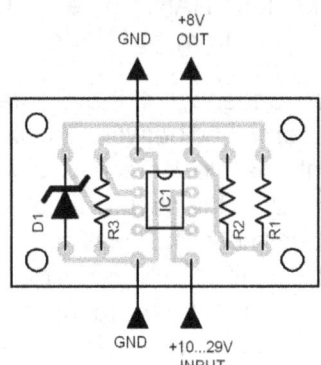

Figure 42.1 Parts Placement Layout

43 AUDIO DELAY LINE

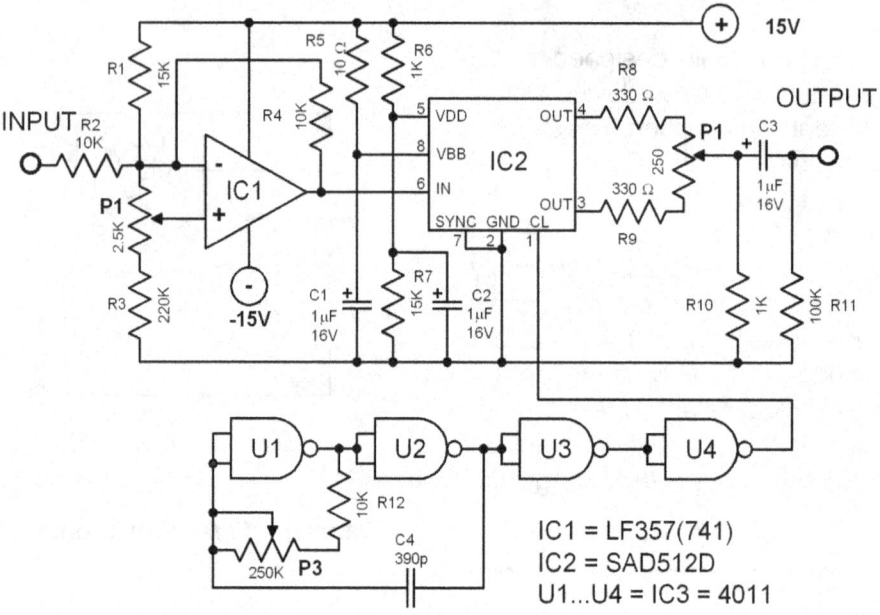

Diagram 43.0 Audio Delay Line

There are different techniques in delaying an audio signal to achieve an echo effect: Hall and echo devices, studio-effect installations, and chamber simulations, etc. The most portable and newest method is the so-called bucket brigade delay (BBD). The delay circuit featured here uses SAD512. This IC has 512 bucket-brigade memory elements with an integrated clock. The analog input signal is converted to a DC voltage by IC1. The four NAND gates U1...U4 function as a clock generator and can be controlled by P3 from 10 kHz to 300 kHz.

The clock frequency of the bucket-brigade memory is controlled between 5 kHz to 50 kHz since the internal clock of the IC is divided into two. The delay of the circuit can be computed through this formula: The delay can be adjusted from 51.2 mS up to 5.1 mS. The maximum frequency of the audio signal that can be processed by the IC is equal to one half of the clock frequency - that is between 2.5 kHz and 25 kHz.

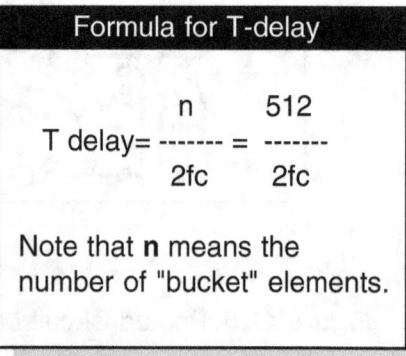

Formula for T-delay

$$T\ delay = \frac{n}{2fc} = \frac{512}{2fc}$$

Note that **n** means the number of "bucket" elements.

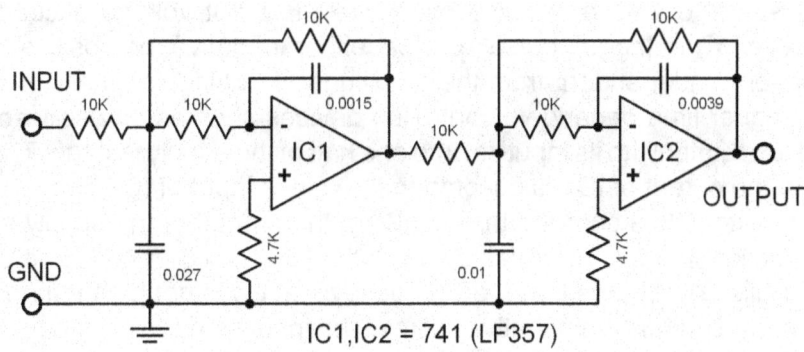

Diagram 43.1 Low-Pass Filter for the Audio Delay Line

The last output signal and the second to the last signal are mixed together by P2 so that the clock frequency can be suppressed as much as possible. Normally, the suppression of the clock signal by P2 is not enough. Therefore the lowpass filter shown in diagram 34.1 must be used to obtain best results. This filter circuit is a fourth order Butterworth filter. The 3 dB cut-off frequency is around 2.5 kHz.

When the frequency of the input signal is higher than one half of the clock's frequency, the circuit will produce unwanted mixproducts. These are the so-called "fold-over" distortions. To avoid this kind of distortion, a second lowpass filter should be used in front of the delay circuit. This filter is identical to the one shown in diagram 43.1.

44 TRIGGER SWITCH

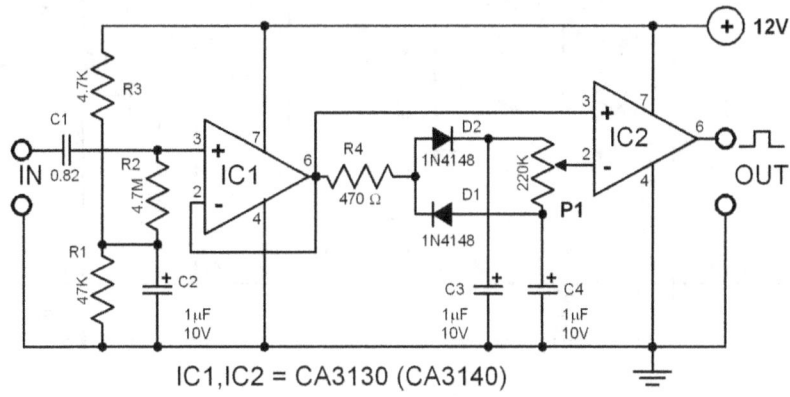

Diagram 44.0 Trigger Switch

Electronic Circuits 1.2

This circuit sends out a short pulse everytime the input voltage exceeds a preset reference level. The trigger pulse can be either negative or positive going. The triggering level is independent from the amplitude of the input signal. This circuit can be used to trigger time dependent/controlled devices such as an oscilloscope.

IC1 is used as a buffer with its input set into one half of the supply voltage. The reference voltage from the output of IC1 is coupled to the minus input of IC2 through D1,D2, C3 and C4. Capacitor C3 is charged through D2 with one half of the supply voltage plus the positive spike of the input signal. Capacitor C4 on the other hand charges through D1 and P1 while D1 secures that the voltage level at C4 is not higher than one-half of the supply voltage minus the negative spike of the input signal. The circuit components are designed so that the reference voltages at the capacitors remain constant at frequencies above 10 Hz.

IC2 works as a comparator. The output voltage from IC1 is coupled directly to its plus input. The reference voltage is set through P1. This level lies between the positive and negative spike levels of the input signal. As long as the input level is higher than the reference level, the output of the circuit is almost equal to the supply voltage. Once the input signal goes down below the reference voltage, the output of the circuit becomes 0V.

45 FREQUENCY MULTIPLIER

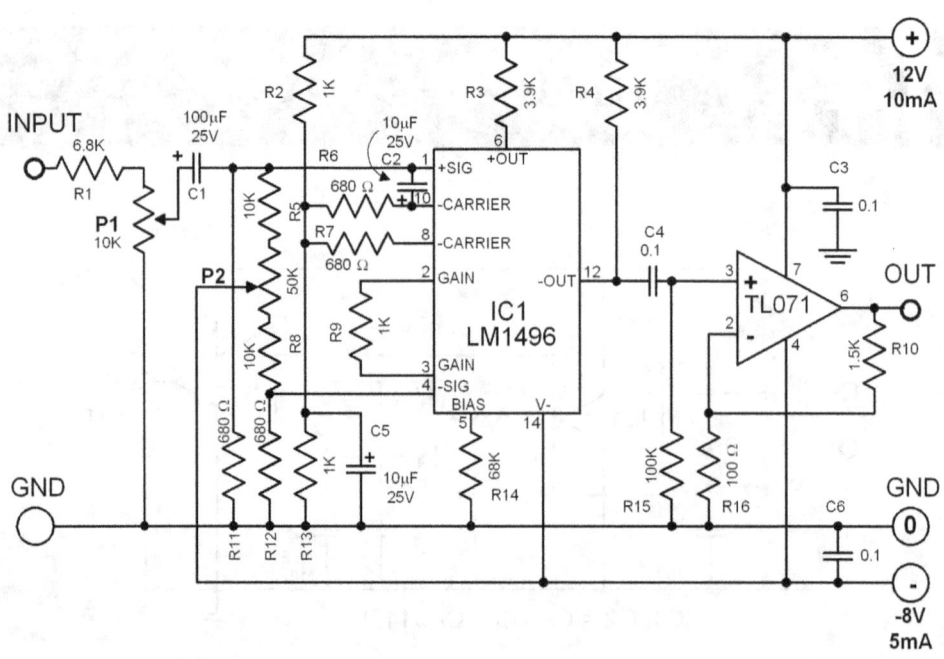

Diagram 45.0 Frequency Multiplier

This circuit doubles the frequency of an input signal. The heart of the circuit is a Modulator/ Demodulator IC LM1496. The IC works according to these trigonometric functions:

$$2\sin(x)\cos(x) = \sin(2x)$$

and

$$\sin^2(x) = 1 - \cos(2x)$$

These functions mean that when two pure sinewave signals of the same frequency are mixed together, the resulting frequency will be doubled.

The frequency doubling is done by IC1. The IC can only process low power signals (25 mV) without distortions. To avoid overdriving the IC, a voltage divider is added into the input circuit. This trims down the input voltage to the allow-able 25 mV. Since this signal is very weak, the output of IC1 is amplified by IC2. IC2 is a non-inverting opamp. The output of IC1 has a DC component of around 8 Volts that is why it is coupled to IC2 through capacitor C4. The op-amp amplifies the signal by about 16 times. The input resistances are set to 680 ohms. The total amplification of the circuit is dependent on the level of the input voltage. If for example the input level is 1.2 V, the amplification is around 1.

If the input level decreases to 0.1V, the amplification becomes 0.1. In order to prevent the input signal from appearing in the output, it is important that the input pins 1 and 4 of IC1 be set by P2 as symmetrical as possible. The basic oscillation can be suppressed up to 70 dB by using an spectrum analyzer. The output signal can be easily distorted since the IC is actually not originally designed for this application.

The distortion factor is dependent on the input level. The signal to noise ratio is between 60 and 80 dB. The current consumption at the positive line is around 10 mA and around 5 mA at the negative line. The phase difference between the input and output signals is around 45 degrees. The positive swing of the output signal is therefore about 45 degrees late.

Parts List:

Resistors:
R1,R14=   6.8K
R2,R9,R13=   1K
R3,R4=   3.9K
R5,R8=   10K
R6,R7,R11,R12=   680Ω
R10=   1.5K
R15=   100K
R16=   100Ω

Capacitors:
C1=   100µF/10V electrolytic
C2,C5=   10µF/25V electrolytic
C3,C4,C6= 0.1µF/50V ceramic

IC's
IC1= LM1496
IC2= TL071

46 ELECTRONIC POTENTIOMETER

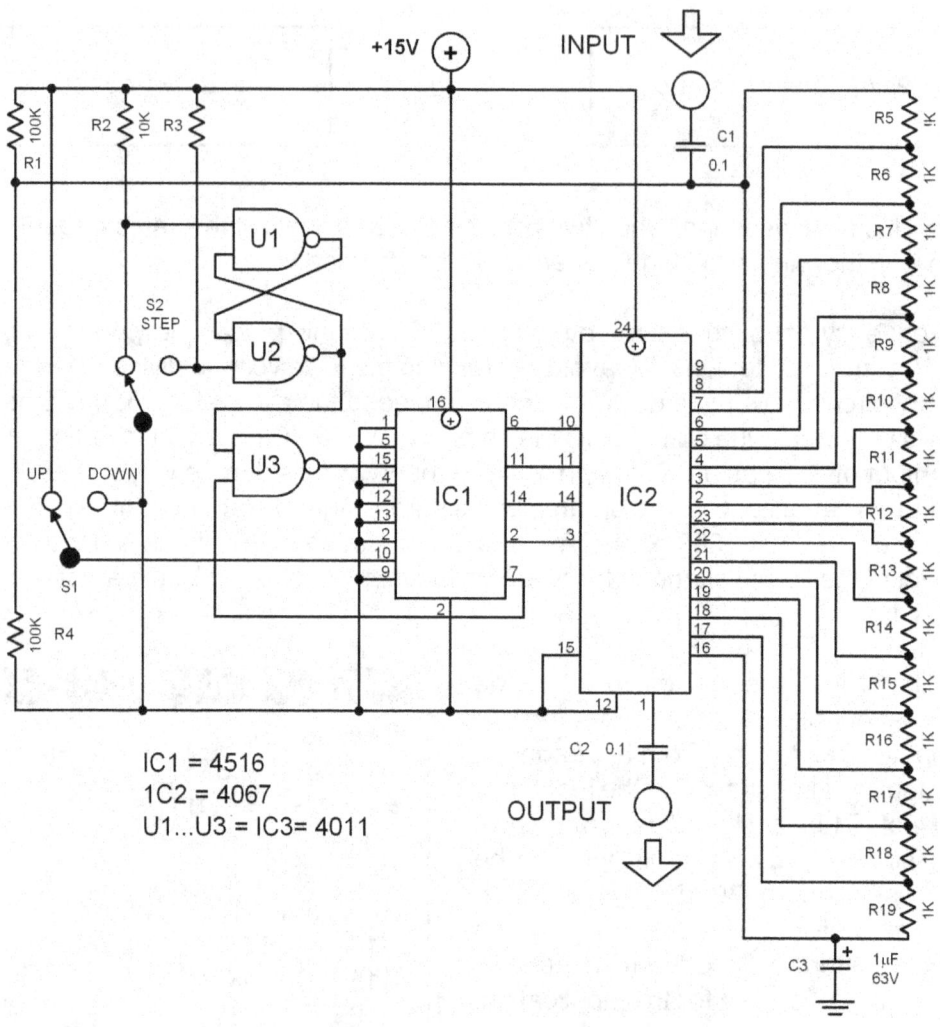

Diagram 46.0   Electronic Potentiometer

This is a digitally controllable potentiometer. The heart of the circuit is a 16 channel analog multiplexer. Controlled by the BCD value at the inputs, the output pin of the multiplexer (pin 1) is connected to any one of the 16 outputs. The output pins are connected with 1k resistors. The IC2 can therefore be considered as a linear potentiometer with 16 steps. The total resistance of this potentiometer is 15k. Other values or characteristics like positive logarithmic can also be achieved by using different resistor values. The setting of the potentiometer is controlled by the counter IC 4516. The position of S1 determines whether the resistance increases or decreases every time S2 is pressed.

U1 and U2 function as anti-keybounce. A logic jump from 000 to 111 or vice-versa is not possible because once the maximum or minimum count is reached, the CO pin (pin 7) of IC2 blocks further pulse entry.

Pin CO is logic 0 when both the counter's setting and pin U/D are logic 0. Once pin U/D is logic 1 and the counter reaches count 15, the output of pin CO reverses to logic 1, and blocks U3. This way, further pulse is prevented since the only way to go further is to count down. The current consumption is less than 1 mA.

47 ELECTRONIC SELECTOR SWITCH

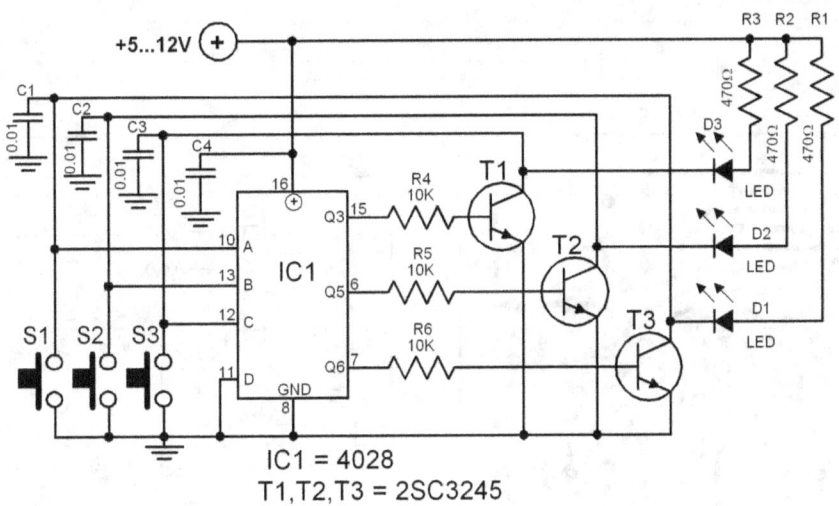

Diagram 47.0   Electronic Selector Switch

Selecting an output signal from several devices to be monitored is usually done through a mechanical switch with several contact points. Ideally, a selector switch secures at first that the old contacts have fully opened before it closes a new pair of contacts. Otherwise two different outputs are connected together creating unwanted results. This ideal switch can be realized by using electronic circuitry like the one featured here. Transistors have the character of being slow in changing from on state to off state that is why using transistors alone does not guarantee the realization of an ideal selector switch.

The circuit featured here uses a CMOS 1-out-of-8 decoder IC. It prevents two transistors from being "on" simultaneously. The collector voltages of the transistors are connected to the input pins of the decoder. For example, if switch S1 is pressed, the decoder will see the binary code 110 and the output 6 will be logic 1 turning on transistor T1. When S2 is pressed on the other hand, the decoder will see the binary code 100(4) and the output 4 will be logic 1. Output 4 is however not used therefore T1 has enough time to completely turn off before T2 will conduct since the output 5 will only be logic 1 when the inputs of the decoder see the binary code 101(5).

The collector voltages of the transistors can control CMOS switches. These CMOS switches then in turn control the actual devices being selected. Relays can also be connected as selector switches.

The switches shown in Figure 47.1 (parts placement layout) are PCB mountable pushbuttons. You can also use other types of momentary contact switches which can be installed remotely.

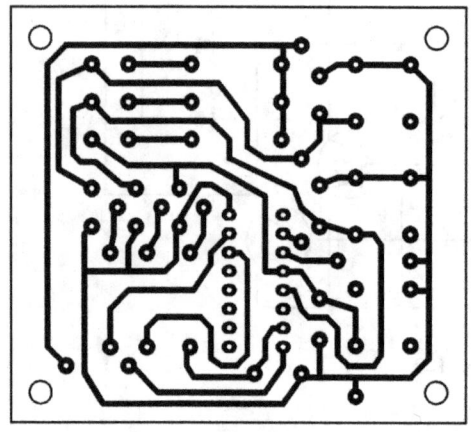

Figure 47.0 Printed Circuit Layout for the Electronic Selector Switch

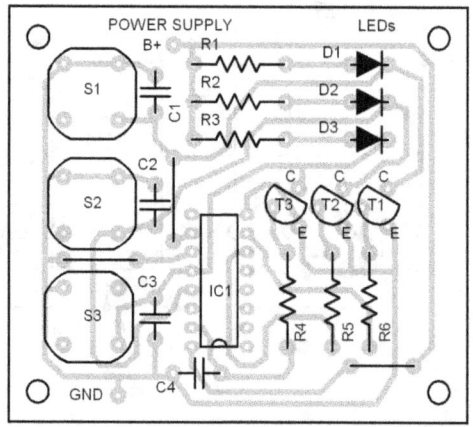

Figure 47.1 Parts Placement for the Electronic Selector Switch

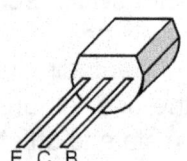

2SC3245

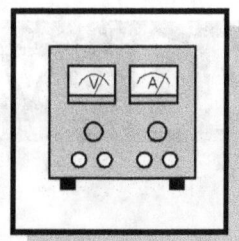

POWER SUPPLIES & CHARGERS

82 Variable Power Supply

83 3-Watt Amp Power Supply

85 0.1V-50V Power Supply

86 Battery Monitor

87 Nicad Battery Charger

90 0-50V/0-2A Power Supply

93 DC Motor Speed Regulator

93 Voltage Monitor

94 Electronic Fuse

95 Current Alarm

98 78XX Regulator Monitor

99 Voltage Converter

100 TTL Power Supply Monitor

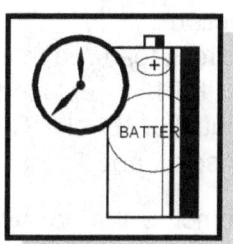

48 VARIABLE POWER SUPPLY

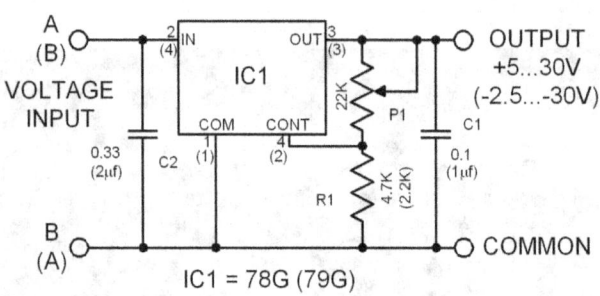

IC1 = 78G (79G)

Diagram 48.0 Variable Power Supply

A stable power supply with an adjustable output voltage from 5 volts to 30 volts can be easily constructed with the regulator ICs UA78G or UA79G. These ICs differ from the common three-terminal regulators (which deliver fix voltage levels only) since their output voltages are adjustable by a voltage level at their control inputs. The maximum current delivered by these ICs is 1 ampere.

The unregulated voltage must be at least 5 volts higher than the desired output level to maintain stability. The input voltage however must not exceed 40 volts. The maximum dissipation of the IC is 15 watts. It has a built-in electrical and thermal overload protection.

The circuit featured in the diagram is designed to give a maximum voltage level of 28 volts. If P1 is replaced with 25K potentiometer, the regulator can deliver up to a maximum of 30 volts. Capacitors C1 and C2 stabilize the IC and they must be connected as close as possible to the IC terminals.

To compute the correct voltage value of the transformer's secondary, use this formula:

where Vts = the voltage output of the transformer's secondary winding and: Vin= is the needed voltage input of the IC.

Formula for Vts
$$Vts = \frac{Vin}{0.7}$$

The IC UA79G delivers negative voltage level. Take note that the two ICs have different terminal connections.

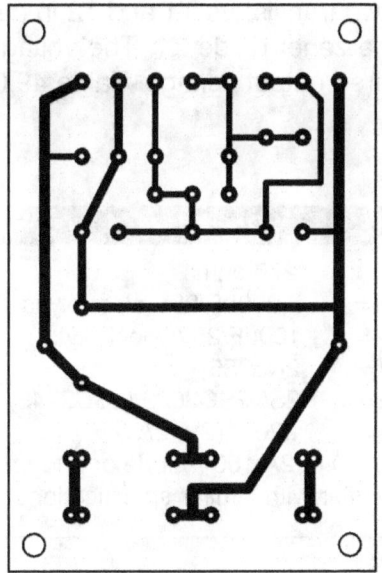

Figure 48.0 Printed Circuit Layout for the Variable Power Supply

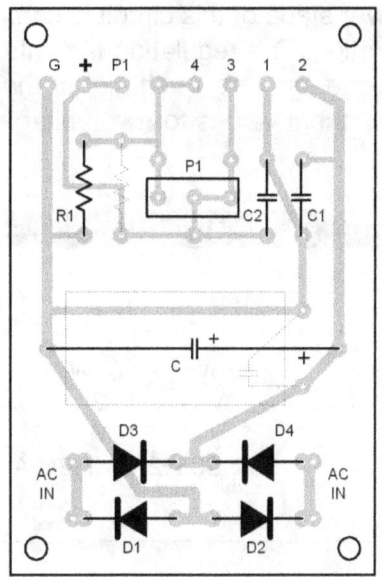

Figure 48.1 Parts Placement for the Variable Power Supply

49 3-WATT AMP POWER SUPPLY

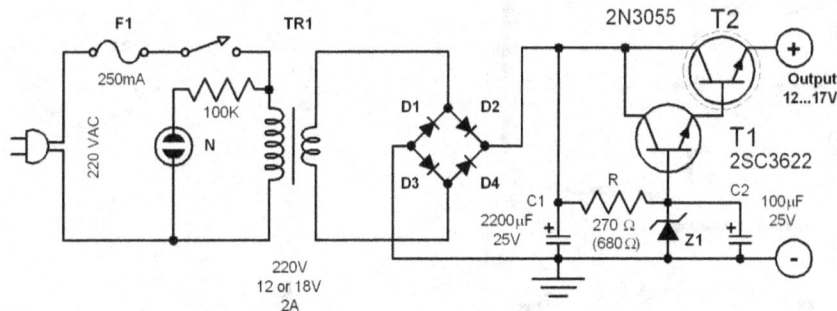

D1...D4 = 1N4001
Z1 = 13V (12V output) = 1N4107
        18V (17V output) = 1N4112
N = 220V Neon bulb

Transistor equivalents:
2SC3622 = 2SC3245, 2SC3248

Diagram 49.0   3-Watt Amplifier Power Supply

The power stage of this circuit is composed of two transistors T1 and T2 in darlington configuration. The regulation is controlled by the zener diode Z1. The voltage of the zener diode must be 1 volt higher than the desired output voltage. Table 49.0 shows the component values for two voltage levels.

Table 49.0		
	12V	**17V**
R1	270W	680W
Z1	13V	18V
Tr1	12V	18V
T2	2SC3622	2SC3245A

Parts List:	
R1=	270 ohms
C1=	2,200µF/25V electrolytic
C2=	100µF/25v electrolytic
T1=	2N3055
T2=	2SC2SC3622 (2SC3245)
Tr=	220V/12V, 2A
D1...D4= 2A/100V diode or bridge rectifier with similar specifications.	

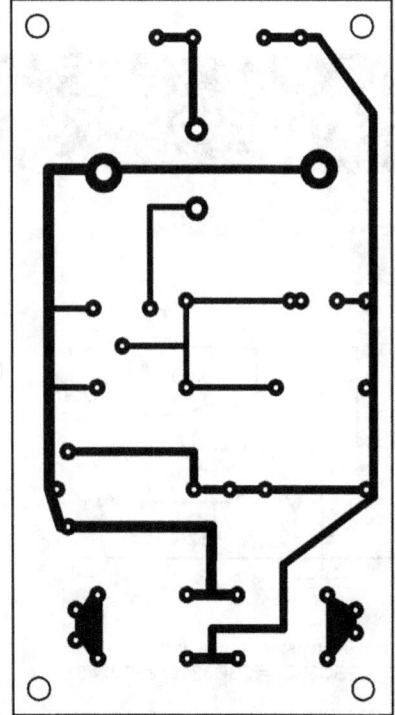

Figure 49.0 Printed Circuit Layout

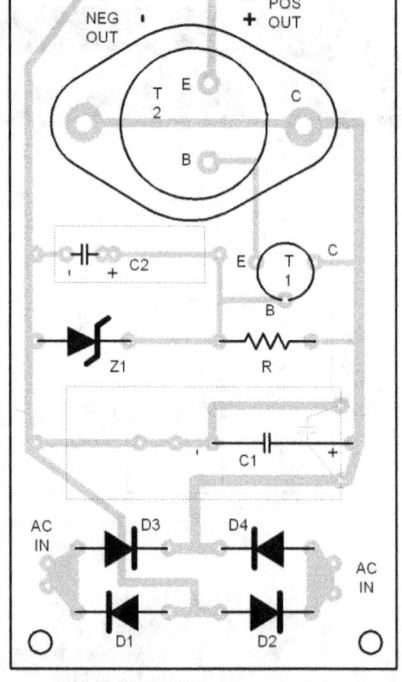

Figure 49.1 Parts Placement

50   0.1V-50V POWER SUPPLY

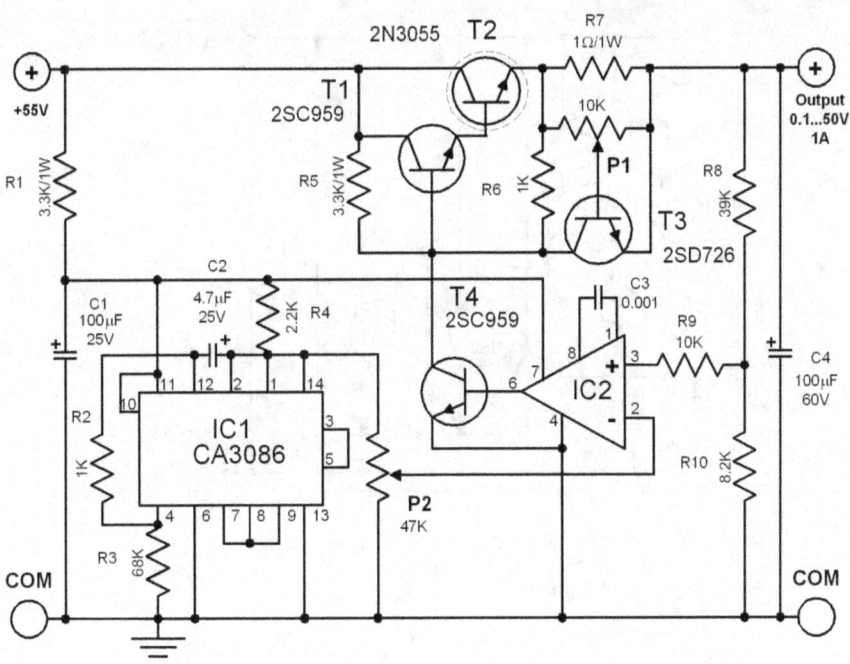

Diagram 50.0   0.1V-50V Power Supply

This power supply circuit is highly stabilized that its output voltage will drop by only 0.005% even though the load changes from 0 to 100 %. Another excellent capability is that the output voltage will change only by 0.01 % if the input voltage fluctuates. The capability of the circuit to be adjusted from 0.1V up to 50V is due to the application of opamp IC CA3130 in the circuit. Transistor T4 raises the output voltage to higher level, and at the same time it separates the lower level opamp from the high level of the output voltage. The reference voltage is supplied by IC1. It is a temperature compensated transistor array with 5 transistors. Four of these transistors are used as reference diodes, and the fifth one sets the output impedance of the reference source.

The reference voltage is set through P1. The opamp CA3130 compares the reference voltage at its minus input to the output voltage at its plus input. The output voltage passes first through a voltage divider before it is fed into the plus input of the opamp. Transistors T1 and T2 work as darlington pair and amplifies the current. Transistor T3 functions as current limiter. The current limit is adjustable through P1, and the lowest current limit is 0.6 ampere. Once potentiometer P1 is set at maximum, current limiting is disabled.

51   BATTERY MONITOR

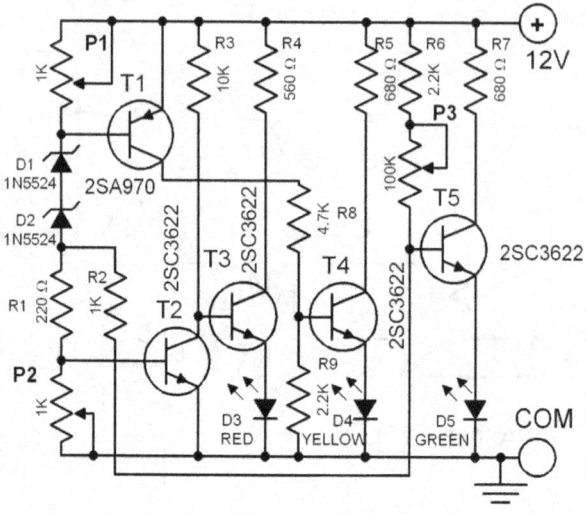

Diagram 51.0   Battery Monitor

A battery's condition can be monitored through the three LEDs of this circuit. The combination of the LEDs' ON condition displays the battery's condition as shown in the following table:

The threshold level of the red LED is controlled by potentiometer P2, yellow LED by P1 and green LED by P3. The setting of each potentiometer is quite critical and must be adjusted repeatedly until the correct threshold level is reached. Finally, the potentiometers must be replaced by fixed value resistors with the appropriate resistance values. These resistance values can be found by measuring the actual resistance setting of each potentiometer.

Table 51.0	
LED On	**Meaning**
D3	less than 12V
D3 + D4	12V - 13V
D4	13V - 14V
D4 + D5	above 14V

52 NICAD BATTERY CHARGER

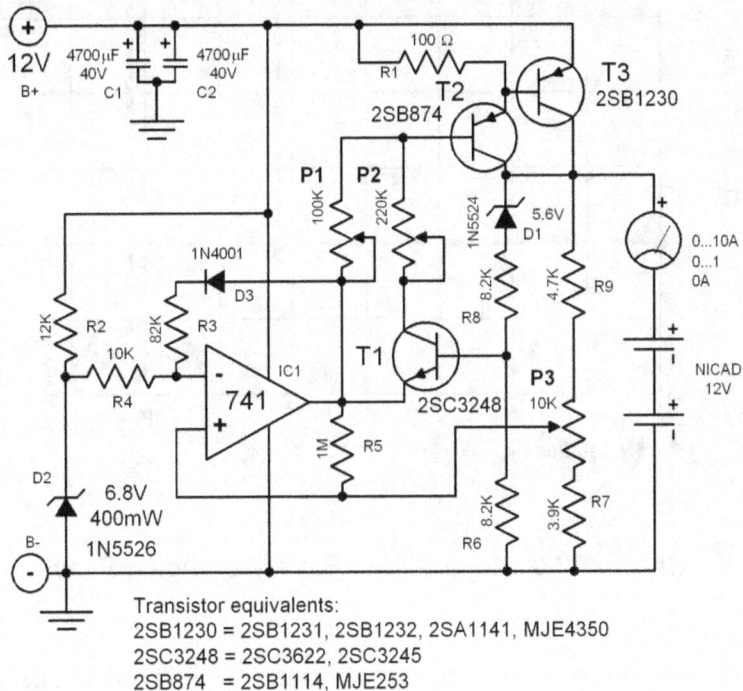

Diagram 52.0 NiCad Battery Charger

Ordinary NiCad chargers simply supply current to the batteries until it is fully charged. It might seem that charging a NiCad battery is a simple straightforward matter. This may be true if we do not consider its bad effect on the life span of the NiCad battery. If we want to use the battery up to its maximum lifespan, we must follow some -quite complicated- techniques in charging it.

The ideal charging method is shown in graph 52.0. The charging goes through three different phases. In phase A-B, the Nicad is charged with a regulated current until it reaches its so-called clamp voltage. Then the current is increased in phase C-D for 5 hours. The ampere-hour of the Nicad is then computed in the "5-hour" value. Once the Nicad voltage reaches about 14.4 V, phase E-F begins where the current is greatly reduced and then sharply decreased until the Nicad voltage reaches about 16 volts. At 16 volts Nicad voltage, the charger automatically switches off.

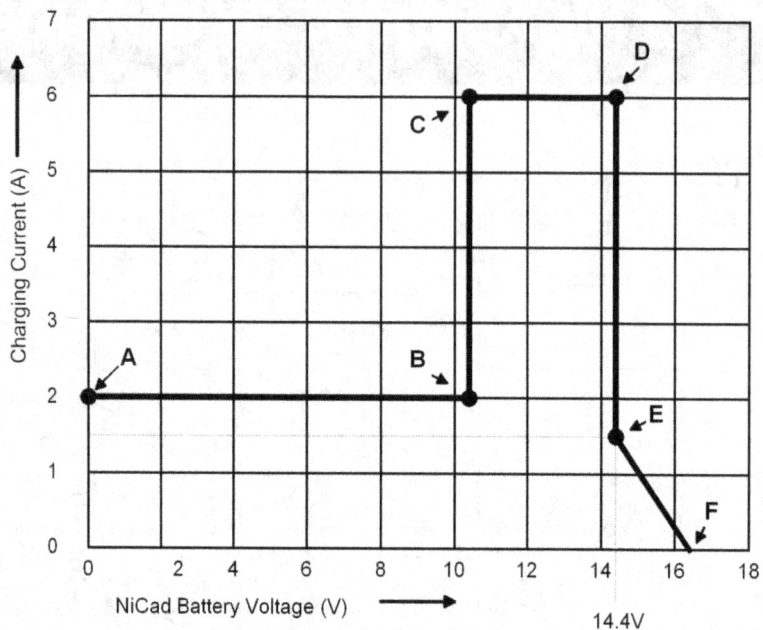

Graph 52.0   Ideal NiCad Battery Charging Phase

The charger circuit featured here approximates the charging phase described. It works in the following way: When the Nicad is still empty -meaning a voltage of 10 volts or less- very little current flows through D1 and transistor T1 does not conduct. IC1 is also shut-off and its output is 0V. In this case the base current of T2, T3 and naturally the charging current depend on the setting of the potentiometer P1. The charging current is then continously regulated to a constant value in phase A-B. Once the Nicad voltage reaches a certain level between 10-14 volts, D3 conducts and switches on T1 starting the C-D phase. At this moment, P1 and P2 control the charging current. P2 is adjusted so that the current will increase during the C-D phase. This charging phase continous with higher current until such a time that the Nicad voltage reaches around 14.4 volts. Once this level is reached, the voltage at the plus input of IC1 increases and at the same time T1 turns off. Since T1 is now off, the exclusive current regulation control is returned to P1. The charging current drops and phase E-F begins. This time however, the charging current is lower than in Phase A-B because the output of IC is now higher. Further increase in the Nicad voltage is channelled through the feedback line composed of diodes D3 and R3 and the current decreases proportionately.

   *Transistors T2 and T3 must be properly heatsinked.*

To align the charger: set P3 so that the output of the opamp is maximum when the Nicad voltage reaches 14.4 volts. When the Nicad voltage exceeds 14.5 volts, adjust p1 so that the charging current drops to the so-called 20-hour-charging value. When the Nicad voltage is between 11 and 14 volts, adjust potentiometer P2 to increase the charging current to the 5-hour-charging value.

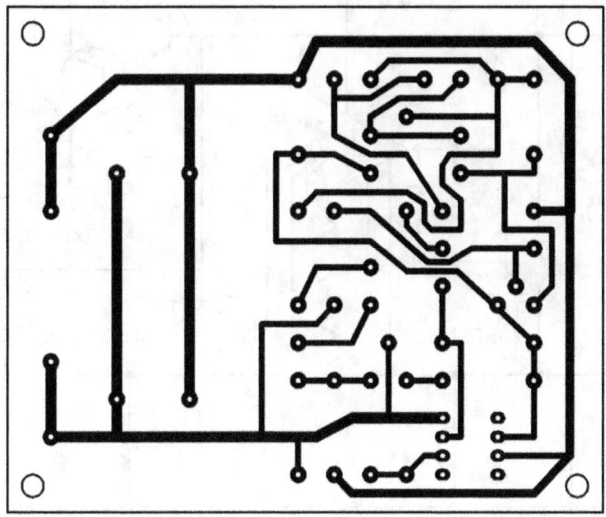

Figure 52.0 Printed Circuit Layout for the NiCad Charger

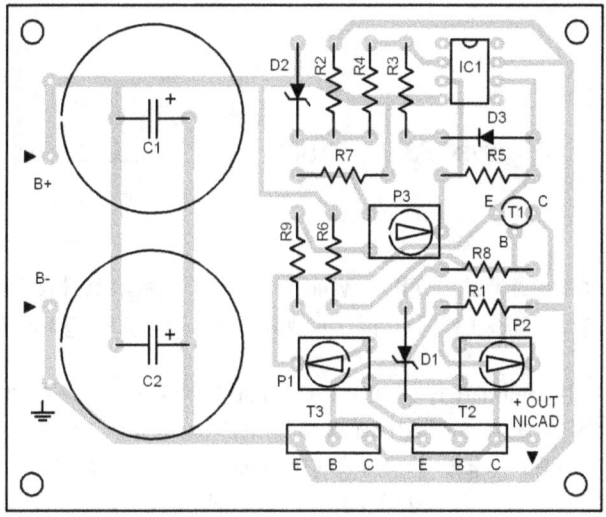

Figure 52.1 Parts Placement Layout for the NiCad Charger

53 0..50V/0..2A POWER SUPPLY

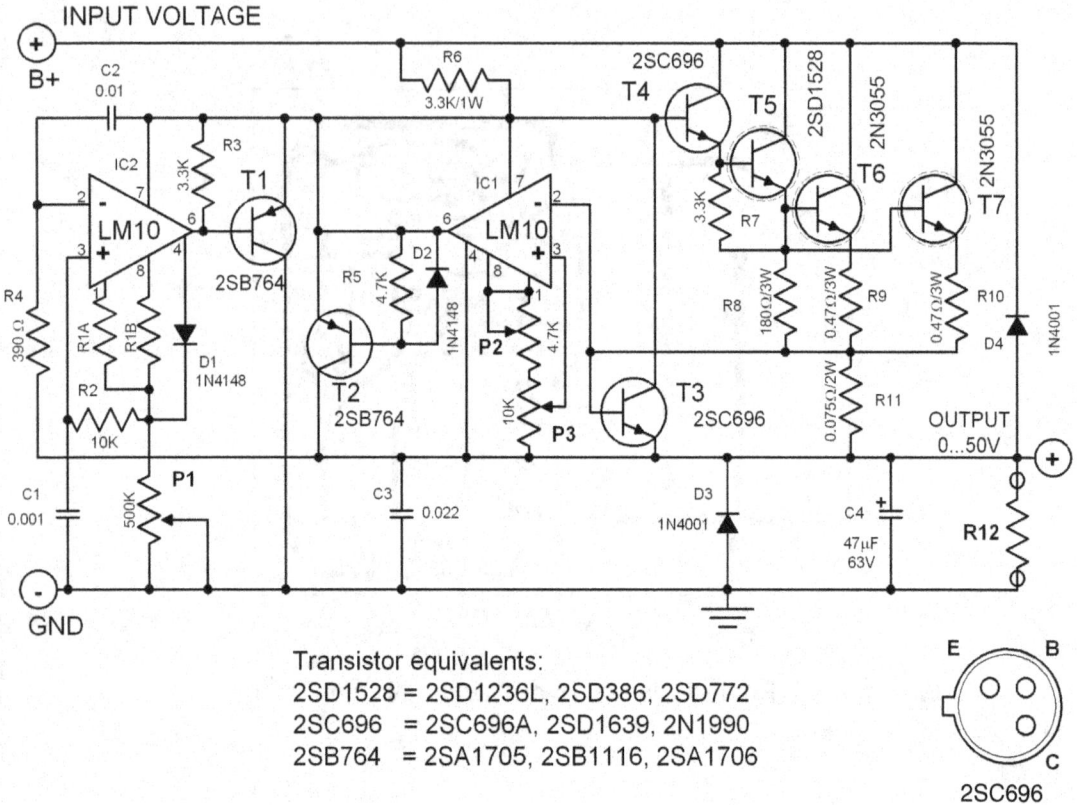

Diagram 53.0 0..50V/0..2A Power Supply

This power supply uses two LM10 ICs which has an integrated reference source. The circuit is designed to be short-circuit proof with variable voltage and current outputs.

The output voltage can be linearly controlled through P1 and the current can be linearly controlled through P3. The maximum current limit is set by P2 and is adjusted only once. The maximum current limit is 2A. The maximum output voltage on the other hand is set by a fixed-value resistor connected in parallel to potentiometer P1. Using a fixed value resistor for this purpose helps in reducing noise at the output voltage.

The voltage regulation works this way: the minus input of IC1 is connected to the output voltage and the plus input is connected to the combination R1/P1. This IC controls the base of transistor T1 so that any voltage difference between the two inputs are equalized. The collector current of this transistor causes a voltage drop at R6 which controls the output voltage through the darlington stage. Pin 1 of LM10 is the reference output. Once the output has stabilized to the desired level, the difference between the two inputs is 0. The voltage level at the junction of R1/P1 is then equal to the voltage level at the minus input of this opamp. Once the reference level is changed by adjusting P1, a voltage difference between the two inputs is sensed by the opamp. The opamp reacts by changing the control voltage until the difference voltage is pulled back to 0. The current is controlled by another LM10. To stabilize the current, a reference level taken from P3 is compared to the voltage at R11. The current flows through this resistor. Since LM10 is not a fast opamp, the current is limited through T3. T3 limits the current to the maximum of 2A.

The minimum output is dependent on the load therefore R12 is added to the circuit as a fixed load. Using 470 ohm for R12 gives a minimum output of 0.4 volts. The maximum output voltage is fixed by R1b and can be set up to 50 volts. The value of R1b used in the circuit sets the maximum output to 45 volts. When it is desired to raise the maximum output to 50 volts, the following components must be replaced:

Transformer secondary 42V/2A, C5= 4,700μF/80V.

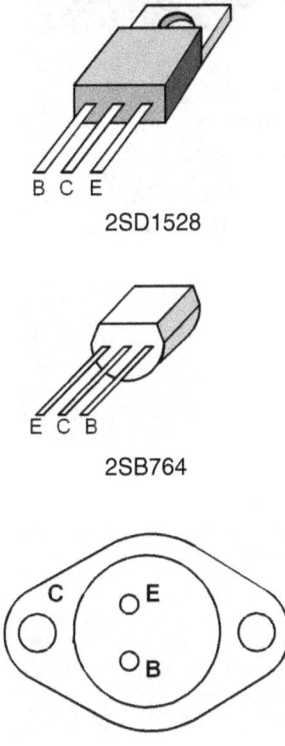

B C E
2SD1528

E C B
2SB764

2N3055

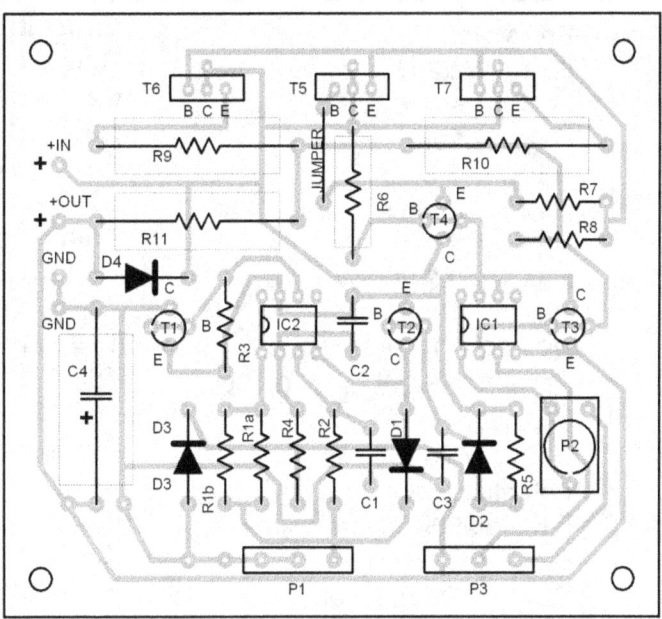

Figure 53.0 Parts Placement Layout

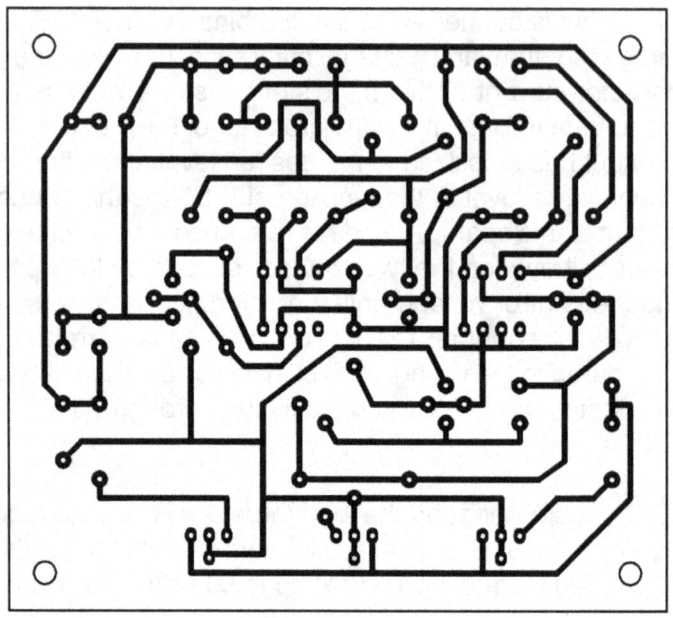

Figure 53.1 Printed Circuit Layout for the 0..50V/2A Power Supply

Parts List:

Resistors:
R1a=   2.2 Ω
R1b=   see text
R2=   10K
R3,R7= 3.3 Ω
R4=   390 Ω
R5=   47K
R6=   3.3 Ω/1 W
R8=   180 Ω
R9,R10= 0.47Ω/3W
R11= 0.075/2W or 2x 0.15Ω
in parallel
R12=   470 Ω/5 W

Potentiometers:
P1=   500K potentiometer
P2=   4.7 Ω trimmer
P3=   10K trimmer

Capacitors:
C1=   0.001 ceramic
C2=   0.01 ceramic
C3=   0.022 ceramic
C4=   47µF/63V electrolytic
C5=   4,700µF/63V electrolytic

Transistors:
T1,T2= 2N290
T3=   BC141
T4=   BC141
T5=   BD241
T6,T7= 2N3055

Diodes:
D1,D2= 1N4148

IC's:
IC1,IC2= LM10
Tr=   220V/36 V(42)/3A

54 DC MOTOR SPEED REGULATOR

This speed regulator uses a single IC LM1014 to control the speed of a DC motor. It senses the increase in the motor-current when the rotation of the motor slows down due to a load. The IC then increases the motor voltage so that the original speed is recovered. Potentiometer P1 varies the speed of the motor.

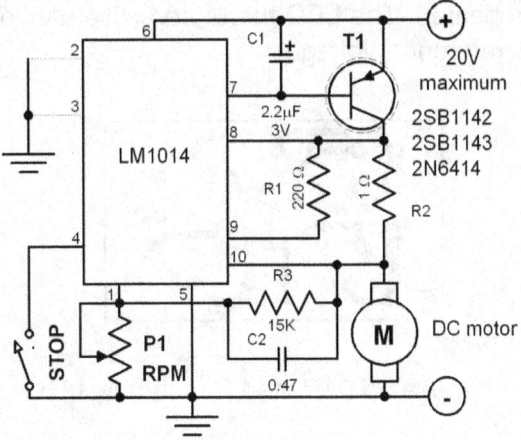

Diagram 54.0 DC Motor Speed Regulator

55 VOLTAGE MONITOR

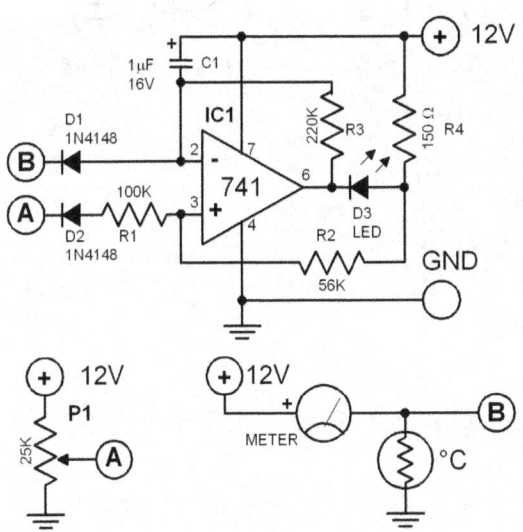

This circuit provides an optical signal through a blinking LED to show whether the voltage level being monitored is lower or higher than a reference level.

The heart of the circuit is a 741 opamp which functions as comparator and oscillator simultaneously. The reference voltage is coupled to point B and the voltage being monitored is coupled to point A. As long as the voltage level at the non-inverting input is higher than at the inverting input, the output of 741 is 12V and the LED is off.

Diagram 55.0 Voltage Monitor

Once the voltage level at point A drops below the reference level, the output of 741 becomes 0 and the LED lights. This time capacitor C1 charges through R3 and the output of the IC1. At a certain charging level, the diode D1 turns off and the capacitor discharges to the minus input of IC1. The IC1 output increases back to 12 V and the LED turns off. The capacitor continous to discharge until D1 turns on again and the process is repeated. The LED therefore blinks as long as the monitored voltage is lower than the reference voltage.

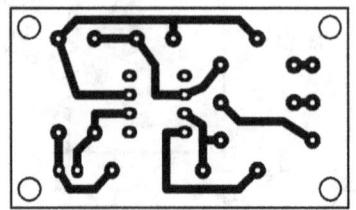

Figure 55.0 Printed Circuit Layout

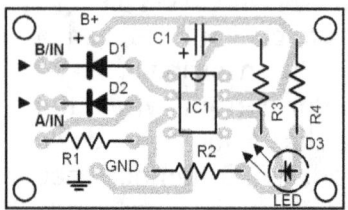

Figure 55.1 Parts Placement Layout

56 ELECTRONIC FUSE

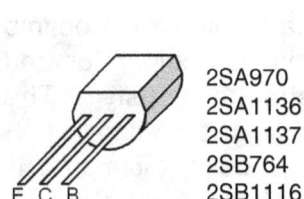

2SA1705
2SA1706
2SA1137

B C E

2SA970
2SA1136
2SA1137
2SB764
2SB1116

E C B

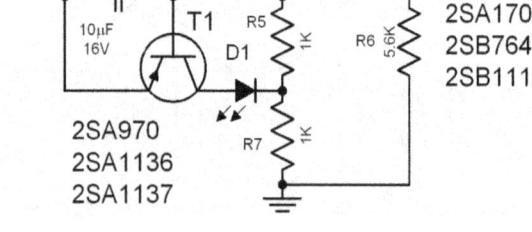

Diagram 56.0 Electronic Fuse

Of course you know what a fuse is and how it functions. This circuit is the electronic version of a fuse. The electronic fuse cuts off the current (like a normal fuse does) and lights up a LED (not with a normal fuse) when a certain current level is exceeded.

The advantage of this fuse in comparison to a conventional one is that it can be reset and re-used (not to mention the cute lighting of the LED). The circuit functions this way: The current flowing to the connected load flows first through R1. Once the voltage drop at R1 reaches 0.5 volts, transistor T1 conducts and switches off T2 and thereby cutting off the current going out of the circuit. The electronic fuse is at this time "blown" and can only be re-used after pressing S1. You then have enough time to check what is wrong with the device connected to the fuse before resetting it.

The value of R1 in the circuit is selected so that the circuit will "blow" when the current is around 500 mA. The circuit cannot deliver currents above 500 mA due to limitations of the transistors used. If you want to change the limit of the circuit change the value of R1 using this formula: $R1 = 0.4(I)$

Maximum output current is 500 mA.

57   CURRENT ALARM

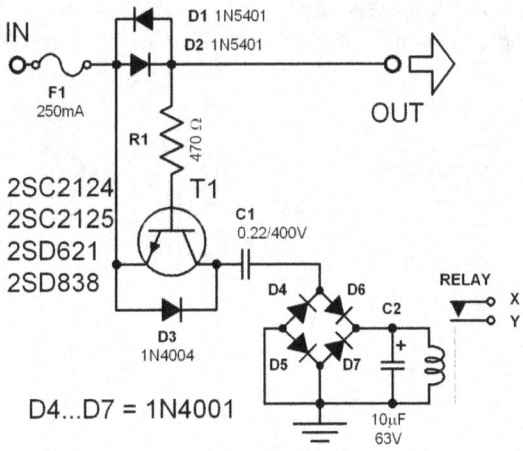

Diagram 57.0   Current Alarm

This circuit is used to monitor the current flow of a device. Once the current flowing through D1 and D2 drops significantly, transistor T1 conducts and the relay closes. The relay then switches on whatever alarm is connected to it.

This circuit has a very wide application area. For example you can monitor exhaust blowers or water pumps. Once the blower or pump motor breaks down, the alarm turns on.

Two PCB layouts are available for this circuit. If you use the first PCB layout, you must install the transistor on a heatsink near the PCB. On the second PCB, you can install the transistor directly on the PCB board. Use a U-shaped heatsink which can be sandwiched between the transistor and the PCB.

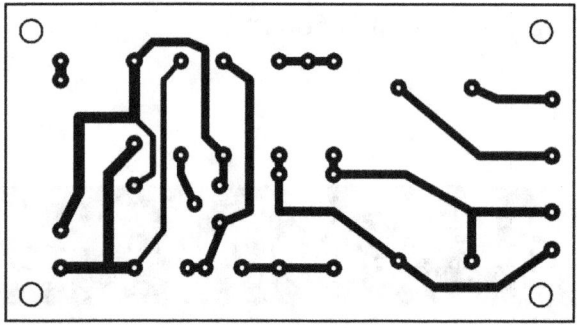

2SC2124
2SC2125
2SD621
2SD838

Note: Bottom view

Figure 57.0 Printed Circuit Layout #1
for the Current Alarm

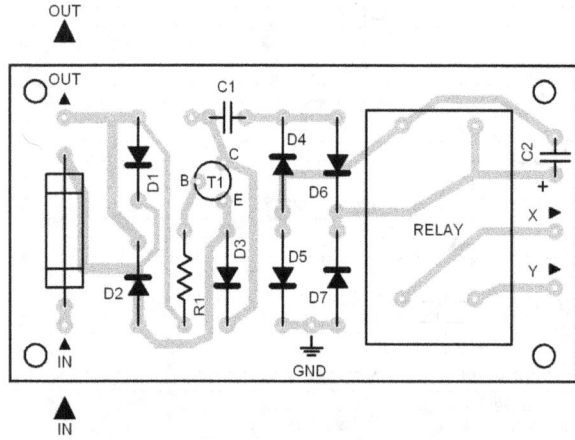

Figure 57.1 Parts Placement Layout #1
for the Current Alarm

Parts List:	
Resistors:	**Transistors:**
R1a= 470 Ω	T1= 2SC2124
Capacitors:	Replacements:
C1= 0.22/400V ceramic	2SC2125, 2SD621, 2SD838
C2= 10µF/63V electrolytic	
Diodes:	
D1,D2= 1N5401 D3,D4,D5,D6,D7 = 1N4004	

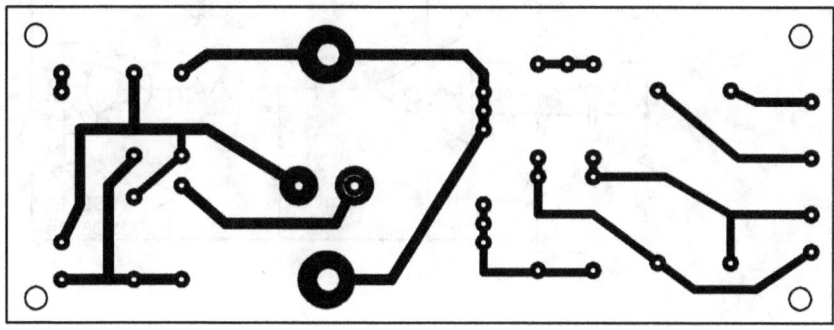

Figure 57.2 Printed Circuit Layout #2 for the Current Alarm

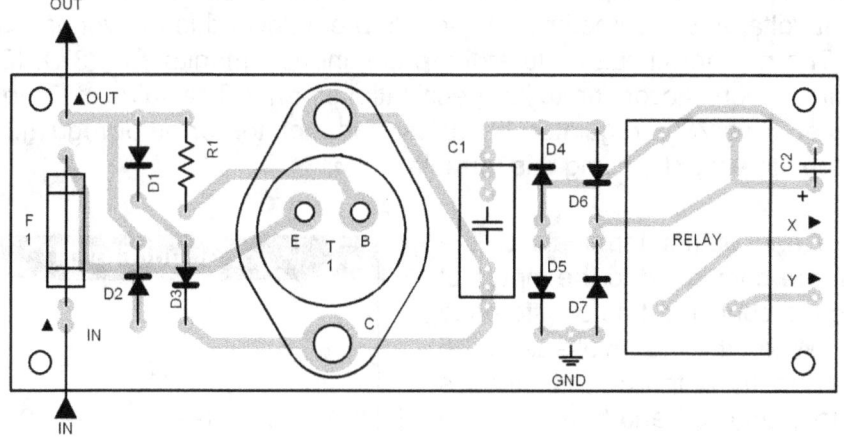

Figure 57.3 Parts Placement Layout #2 for the Current Alarm

58 78XX REGULATOR MONITOR

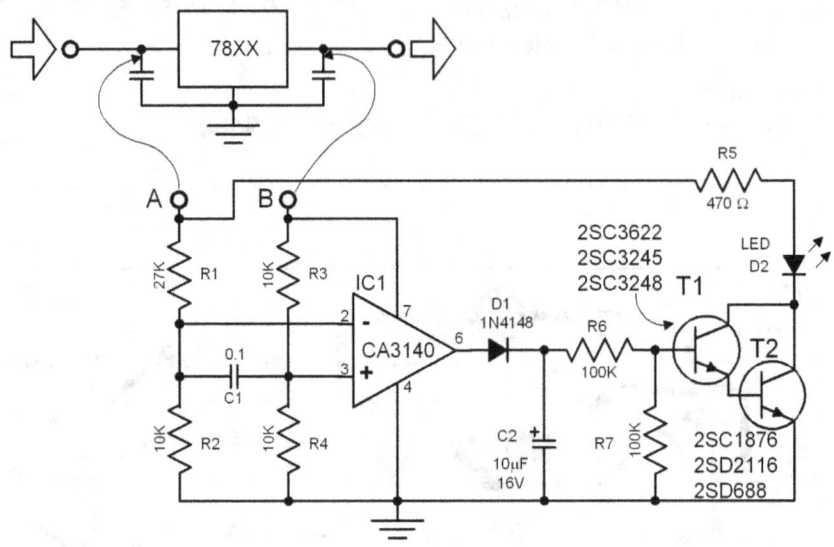

Diagram 58.0 78XX Regulator Monitor

Although 78XX 3-terminal regulators are known to be stable, problems sometimes arise in their applications. The input voltage of these ICs must be at least 3 volts higher than the output voltage. But sometimes these ICs breakdown due to overloads or internal defects. The monitor circuit featured here continously monitors a 78XX IC regulator whether if functions according to its specifications or not. The value of R1 in the circuit is computed for a 7805 regulator. If you use the monitor for other regulator IC types, change the R1 value by using this formula:

The input and output terminals of the regulator are connected to the inputs of IC1 which is configured as a difference amplifier. When the input voltage of the 78XX IC is low, the output of IC1 increases enough to charge C2 and turn on T1. In this case, the LED lights up. C2 holds the LED for about 10 mS longer before it finally turns off so that even short voltage collapses can be registered.

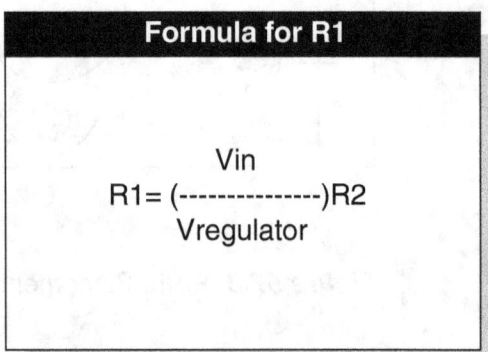

Formula for R1

$$R1 = \left(\frac{Vin}{Vregulator} \right) R2$$

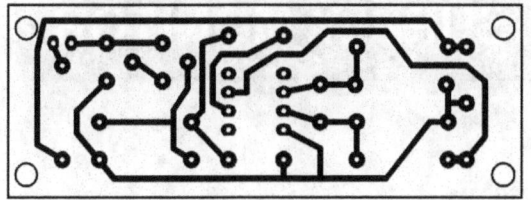

Figure 58.0 Printed Circuit Layout for the 78XX Regulator Monitor

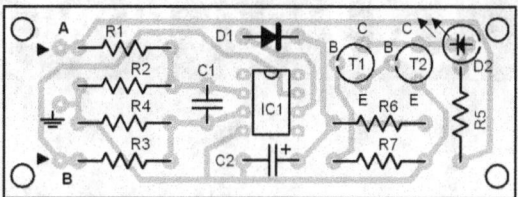

Figure 58.1 Parts Placement

59 VOLTAGE CONVERTER

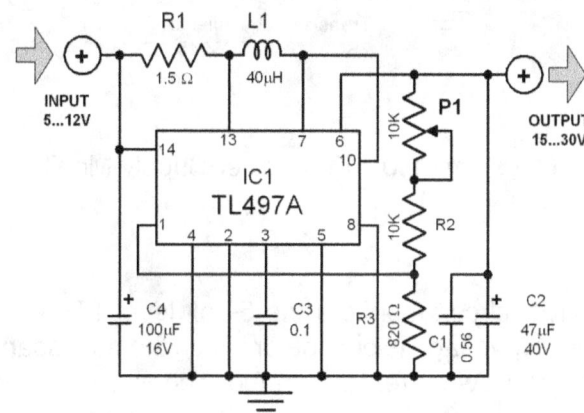

Diagram 59.0 Voltage Converter

This circuit converts an input voltage of 5...12 volts to a higher level of 15...30 volts. This is specially helpful in mobile applications where power supply levels are commonly limited to 12 volts supplied by batteries.

The circuit uses the IC L4973 as a flyback-voltage converter. The choke coil is 40µH/2A. Capacitors C2 and C1 suppress voltage spikes. The maximum output current depends on the difference between the input and output voltages. This maximum is around 100 mA. The ripple voltage is relatively low. The standby current is around 8 mA and the efficiency is about 70%.

The maximum output current is 100mA!

60 TTL POWER SUPPLY MONITOR

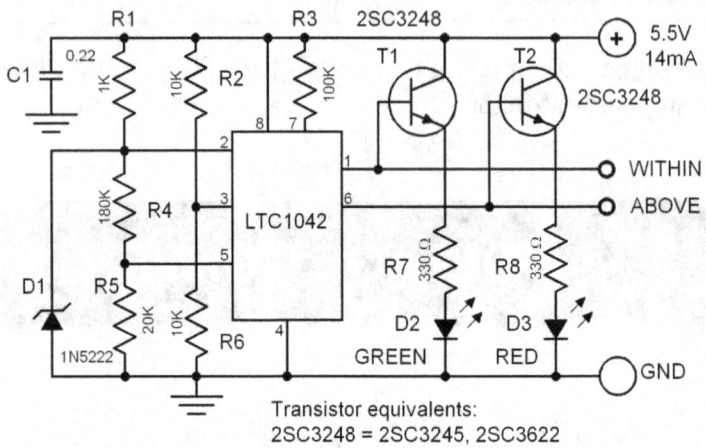

Diagram 60.0 TTL Power Supply Monitor

This simple to construct circuit monitors the 5-volt level TTL power line and gives a signal whether the supply voltage is outside or within the necessary range or "window". The heart of the circuit is a low-current integrated window comparator. The center of this window is set at 2.5 volt +/- 0.005 volt by the bandgap reference diode D1 which is connected to pin 2 of LTC1042. The width of this window must be 20% (+/- 10%) of the reference voltage.

The reference voltage is reduced by 25% through resistors R4 and R5 and fed to pin 5 (width/2 input) of the IC. The monitored voltage is then fed to pin 2 (window center input) so that the green LED (D2) will light up when the voltage is within the desired range. Otherwise, the red LED (D3) will light up signalling that the voltage is out of range.

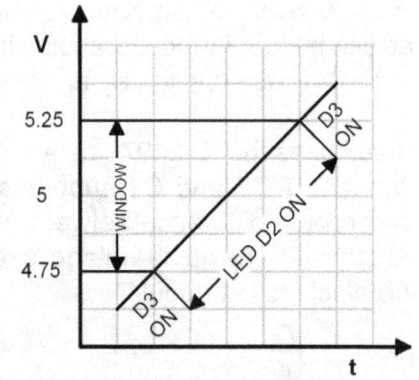

Table 60.0 Time vs voltage

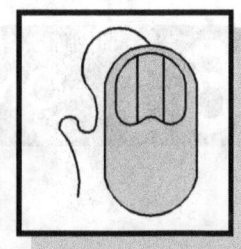

DIGITAL & COMPUTERS

102 One-Chip Baud Generator

103 RS-232 Switch

104 A/D Joystick Converter

105 Serial Data Converter

106 Keyboard Doubler

108 Printer Computer Switch

109 Serial A/D Converter

110 8-Bit A/D Converter

111 Parallel-Serial Converter

112 Video Signal Mixer

113 TTL Squarewave Generator

114 CPU Clock Generator

61 ONE-CHIP BAUD GENERATOR

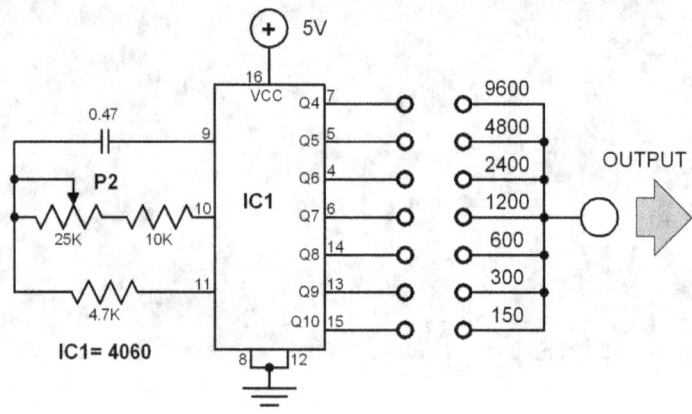

Diagram 61.0 One-Chip Baud Generator

This circuit offers a cheap alternative for generating clock signals with 7 different baud-rates. Baud rate generators are often used in serial I/O of computers. It happens very often that the same generator module cannot be used for both the "transmitter" and the "receiver". This circuit is highly invaluable in such cases. It provides an extra baud rate generator.

The generator circuit is composed of one CMOS-IC, two resistors, 1 trimmer and a capacitor. The CMOS IC 4060 is a 14 stage binary counter with an internal oscillator. Its oscillator frequency can be set by external R/C components. The oscillator output is connected to the clock input of the counter. The IC begins to count upward immediately after the power is supplied since its reset pin is connected to ground. The clock signals can be taken from the outputs. The higher the pin number, the lower is the signal frequency.

The output signal can be selected by either soldering a jumper wire between the proper output pin and the main output terminal or installing a PCB DIL switch. The oscillator frequency can be fine tuned by potentiometer P1.

62   RS-232 SWITCH

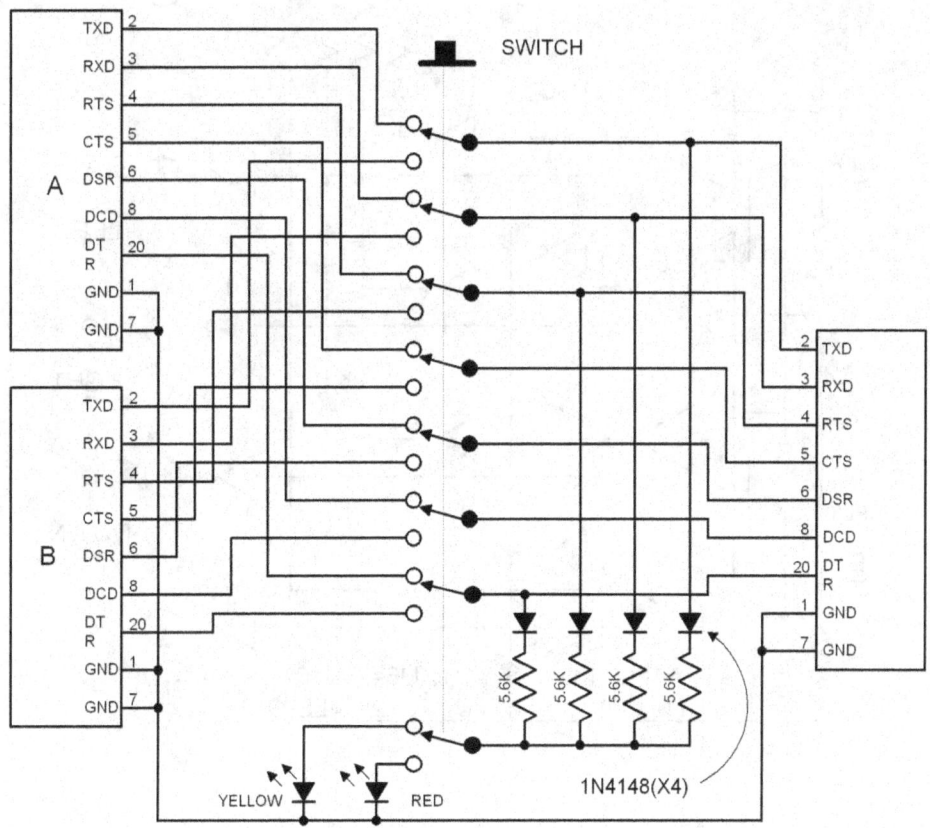

Diagram 62.0   RS-232 Switch

This switch circuit makes it possible for you to re-route the RS232 terminal of your computer without breaking your back from continously bending to reach the backpanel of the devices. The circuit is very simple to build and has an LED display which lights when data are being transmitted through the lines. Once the LED is lighted, you'd better not move the switch at this moment. The connection data of the circuit is for type D connectors with 25 pins.

63 A/D JOYSTICK CONVERTER

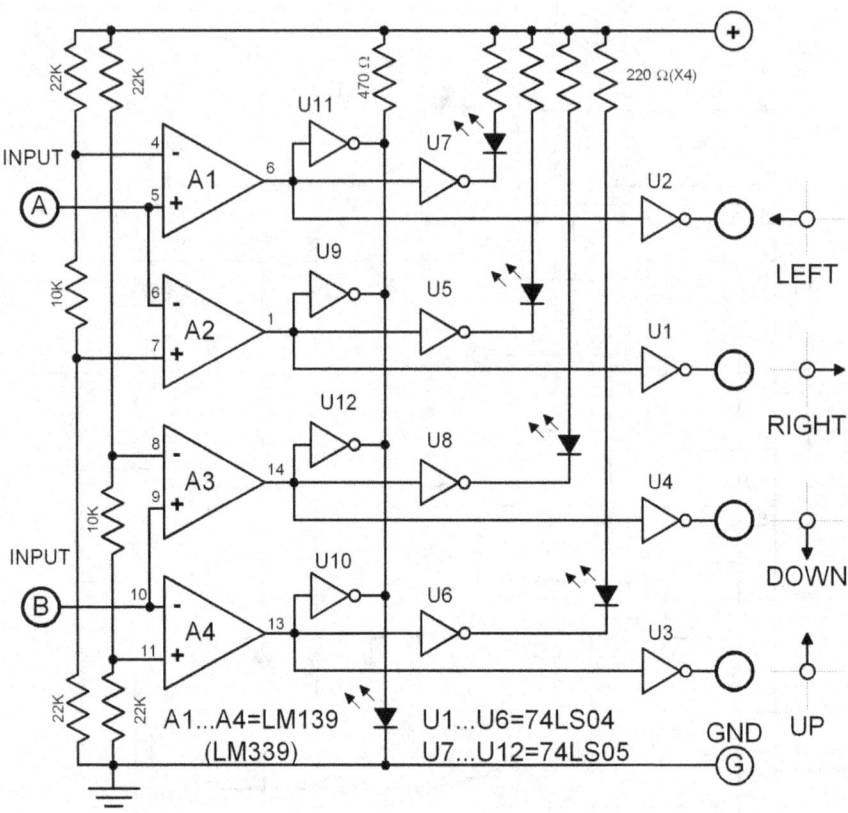

Diagram 63.0 A/D Joystick Converter

Joysticks can be divided into two types: the one with potentiometers (analog) and the one with contact switches (digital). Unfortunately, the potentiometer variants cannot be used for computers which accept only contact-switch joysticks. The Commodore C64 computer is an example. Since this computer is widely used, the circuit here helps to convert the potentiometer type joystick into a switch type one.

The analog joystick is connected to points A and B and to the supply line. The actual switches are composed of 4 comparators in IC1. The comparators convert the direction information of the joystick and passes it on to the computer. The outputs of the comparators are buffered by U2...U3. The remaining inverter gates are responsible for controlling the LEDs. The LEDs show the direction of the jostick. The inverter gates U9...U12 are NOR-wired connected so that LED5 will only light when all comparator outputs are 0. The current consumption of the circuit is around 25mA.

64 SERIAL DATA CONVERTER

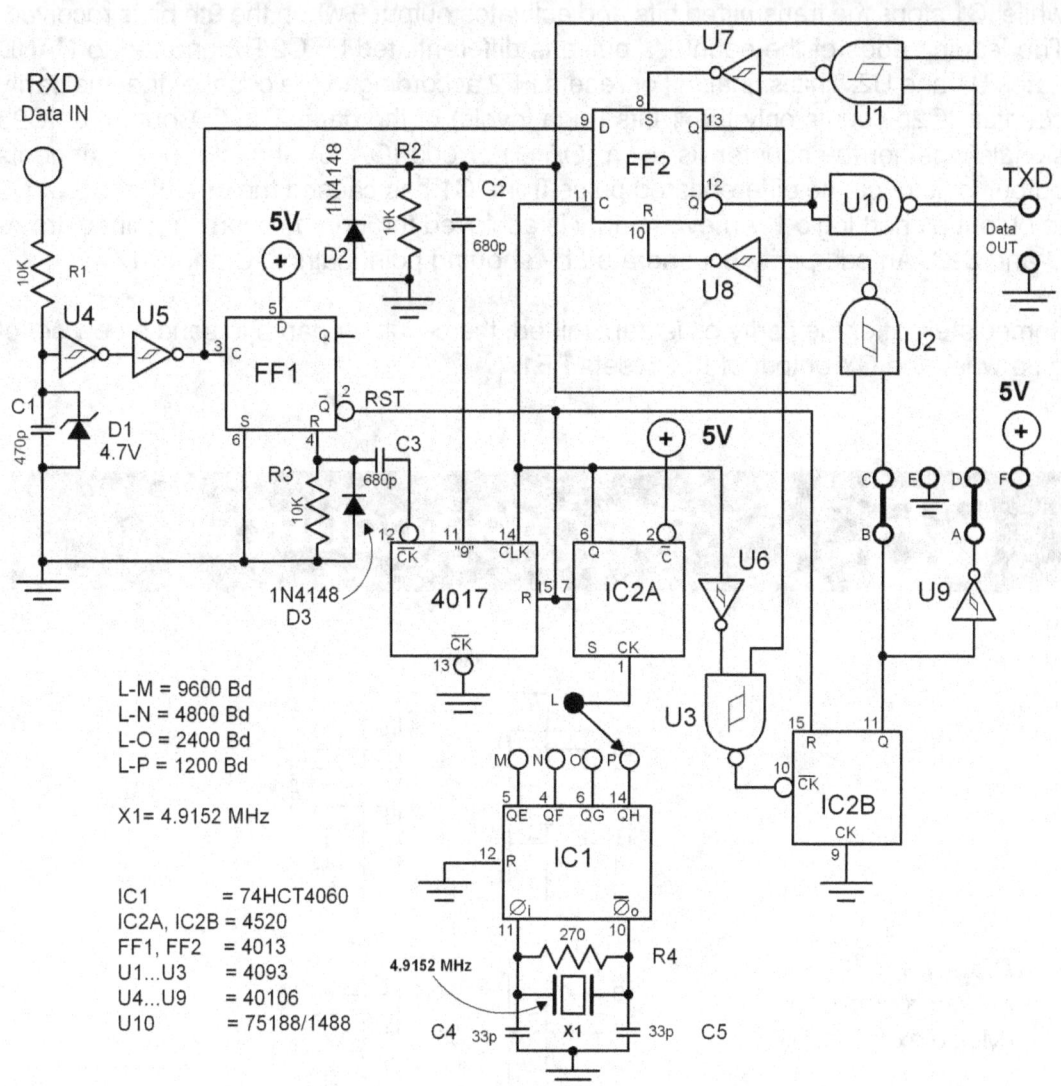

Diagram 64.0 Serial Data Converter

By using this converter circuit the computer can generate a data format with 1 startbit, 8 databits, no parity bit, and 1 stop bit. This converter can generate both even and odd parity bit without the help of any handshake routine. The circuit has its own baudrate generator with a variable baudrate.

The leading edge of the startbit causes the output Q of FF1 to become logic 0 releasing the counters IC2a and IC2b to start counting. This time ICa counts the clock pulse coming from baudrate generator IC6. The frequency of this signal is 16 times higher than the bitrate at the I/O line. Flip-flop FF2 and IC are clocked at CK by a signal equal to the pulse rate of the databits. The start bit of the following 7 bits are passed on to FF2 while IC1 stops the transmitted bits and activates output 9 when the 9th bit is received. The leading edge of the counter's pulse is differentiated by C2/R2 and fed to NAND gates U1 and U2. This signal set or resets FF2 according to the count of the the parity counter. IC2b counts only the 1 bits (high levels) of the data. The OA output of IC2b signals whether the counter is even (OA=1) or odd (OA=0) and lets FF1 change its output logic once the differentiated pulse from IC1 has caused the output of U8 or U9 to output a short logic 1. An even parity is achieved through by shorting point pairs A-D and D-C. An odd parity is generated by shorting point pairs A-C and B-D.

Immediately after the parity bit is transmitted, the circuit prepares to send a new set of data when the CY output of IC1 resets FF1.

65 KEYBOARD DOUBLER

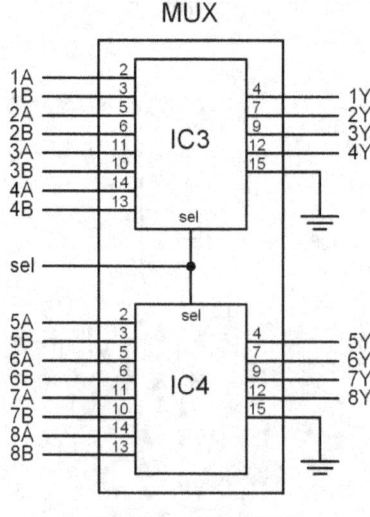

MUX

IC3, IC4 = 74LS157

Diagram 65.0
Keyboard Doubler
(Multiplexer Module)

This is a multiplexer circuit which is applied as a doubler for Apple keyboards. It enables an extra keyboard to be operated in parallel to the system keyboard. In principle, it is a simple electronic switch labeled as MUX in the circuit. Both inputs of the MUX are connected to the data lines of the keyboards.

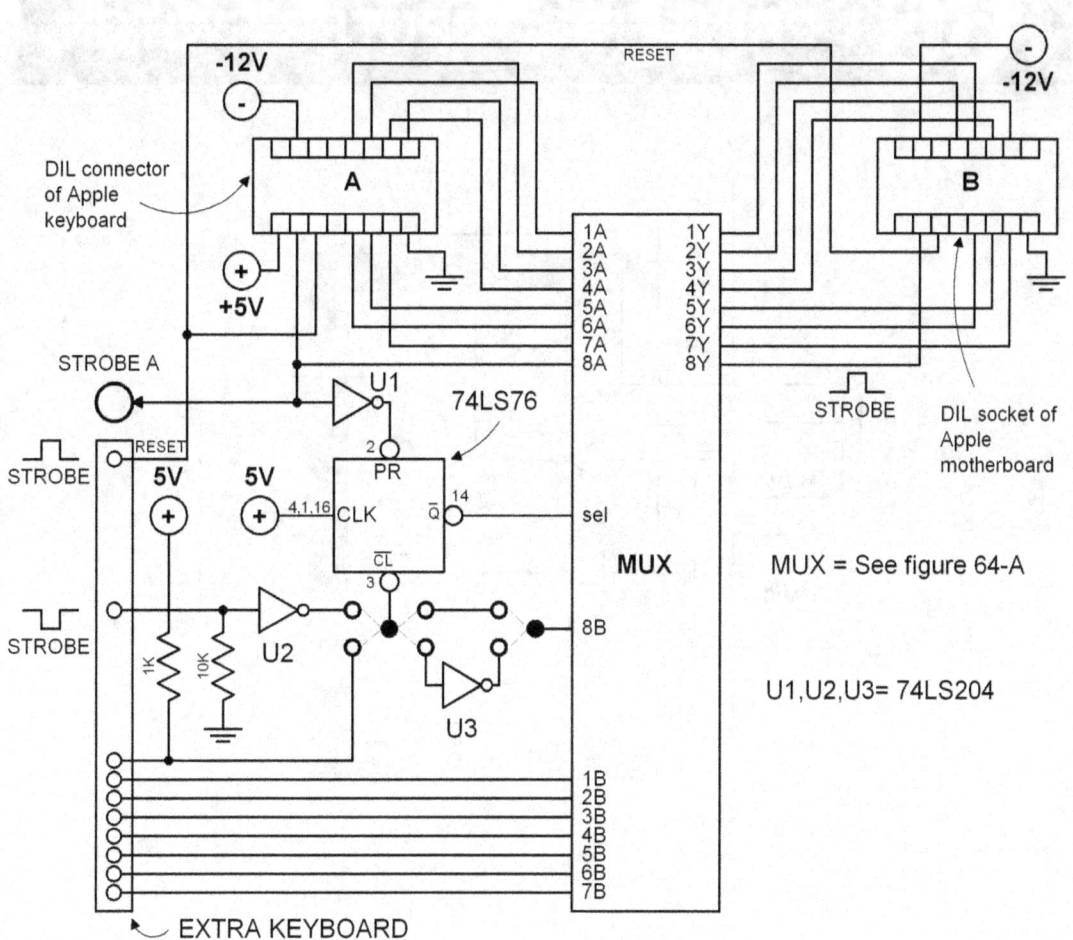

Diagram 65.1 Keyboard Doubler

The MUX automatically connects the right keyboard to the computer. This auto-switch is controlled by the keyboards themselves. Everytime a key is pressed, the keyboard generates a strobe pulse along with the data bits. This strobe pulse then sets or resets the flip-flop IC2 depending on which keyboard has generated the strobe pulse. This pulse serves as a select-signal for the MUX. The multiplexer MUX is built from two 74LS157 ICs. Each IC has four 2-to-1 multiplexer so that it can accomodate all 8 inputs. When the select input of both ICs is logic 0, the outputs 1Y to 8Y are connected to the inputs 1A to 8A. When the logic is 1, the B inputs are connected to the output lines.

66 PRINTER-COMPUTER SWITCH

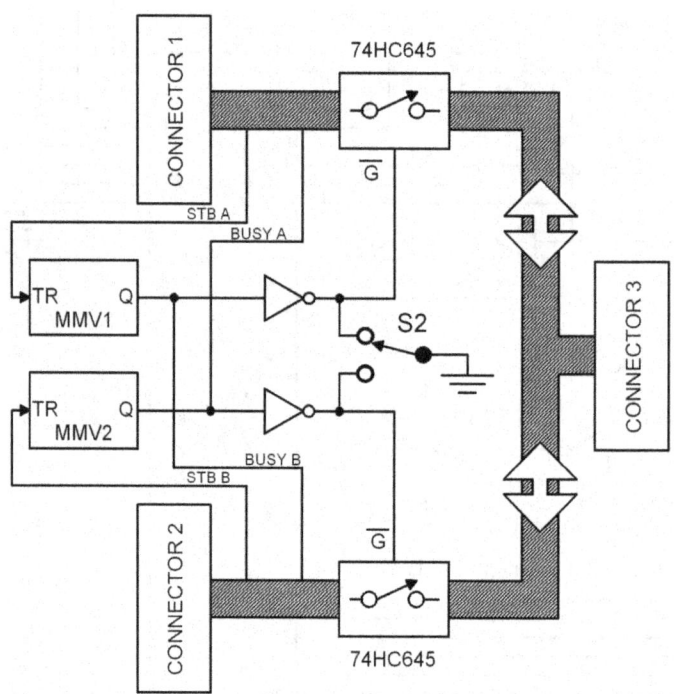

Diagram 66.0 Printer-Computer Switch

With this circuit you can connect either two printers to a single computer or two computers to a single printer. It guarantees a smooth switch over and secured data transfer. The circuit works according to the first come first serve principle. The block diagram shows how it functions. For example if computer 1 wants to print, STBa becomes active and triggers the resettable monoflop Multivibrator (MMV1) . The output of MMV1 goes to logic 1 and closes the elecronic switch thereby preventing the activation of line B by sending a busy-line-signal to computer 2. As long as computer 1 prints, the circuit remains at this condition until the STBa becomes inactive. MMV1 resets only 30 seconds later after STBa became inactive. The two LEDs show which printer or computer is active at the moment.

However, the circuit may cause problems by long programs which process printable data more than 30 seconds later after the PRINT command is received. Changing the delay time of the monoflop by replacing the value of capacitor C1 is therefore necessary in such a case.

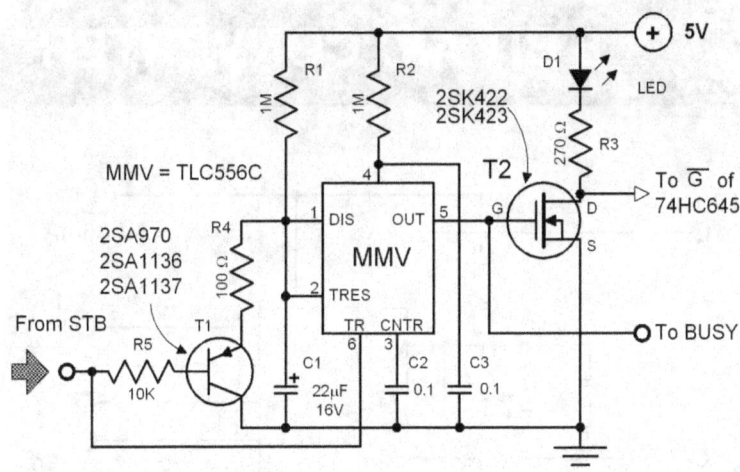

Diagram 66.1 Printer-Computer Switch

A simpler method is to use a mechanical switch to control the ICs. This will, however, defeat the automatic feature of the circuit. Switch S1 serves as reset for emergency cases. The circuit consumes only around 50 mA.

67 SERIAL A/D CONVERTER

This A/D converter uses a TLC548 chip from Texas Instruments which is designed as a complete single-chip A/D converter. The converter chip serially delivers the converted data into the computer. The transfer of data into the computer flows through 3 lines. When a control signal from the computer switches the CS pin of the converter chip, the data bits at pin 7 is delivered to the computer. Since the output of the chip is tri-state TTL compatible, it is possible to connect several ICs in parallel. The data conversion takes about 17mS. The internal clock of the converter and the clock at its I/O CLK input are closely dependent with each other.

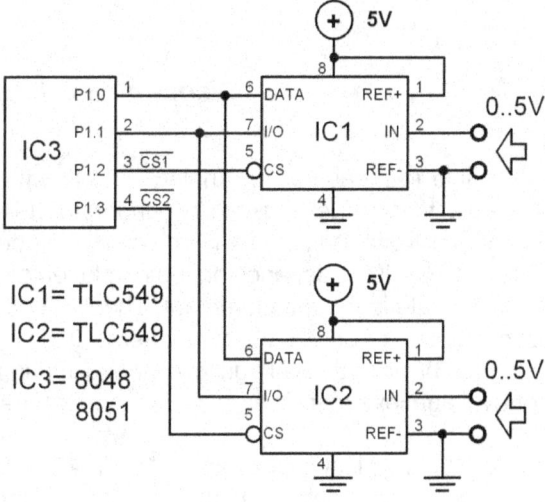

Diagram 67.0 Serial A/D Converter

68 8 BIT A/D CONVERTER

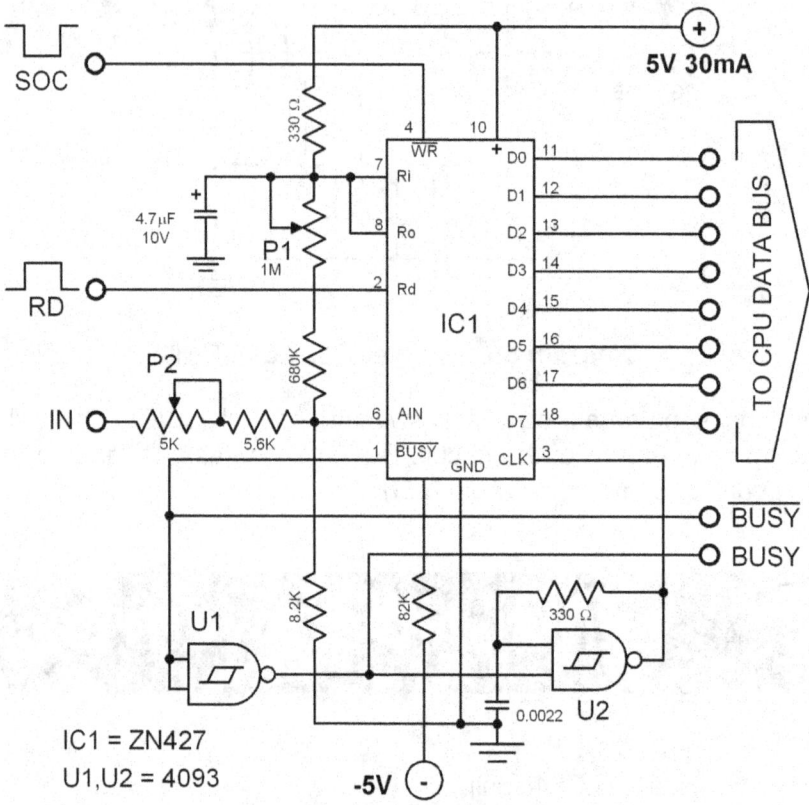

Diagram 68.0 8 Bit A/D Converter

To be able to measure an analog voltage with a computer, the analog value must first be converted to a value that can be understood by the computer. Since computers work and understand digital only, the analog value must be converted to a digital value with the desired precision. Precision depends on the resolution of the data. That means: the more bits used, the more accurate is the measurement. The A/D converter here serves for that purpose. Although it is composed of only a few components, it is versatile, fast and precise. The maximum input voltage is 5V. With this voltage level, the resolution is around 19.6 volts per sampling step. Other input voltage levels are also possible to measure if the input divider is changed.

The IC has a typical conversion of 10nS. AC voltages can be measured -meaning digitalized- and processed in machine language. A logical 0 at the -WR input starts the conversion cycle and the busy line becomes logic 0. The schmitt trigger U1 is triggered and generates a clock frequency of 900 kHz. When the conversion is finished, the -Busy line goes back to 1 and the CPU can now read the 8 bit data from the memory of the A/D chip.

The read and start signals must be decoded to a CPU compatible signal. The calibration is done through P1 and P2. A computer program is necessary to calibrate the circuit. Such a program is not difficult to write. First P1 is set to 0V input voltage so that the computer reads 0. P2 is then set to the maximum voltage level so that FF_{hex} is read by the computer. The circuit is linearly calibrated if the computer reads 125 (80_{hex}) when the input is 2.5 volts.

69 PARALLEL-SERIAL CONVERTER

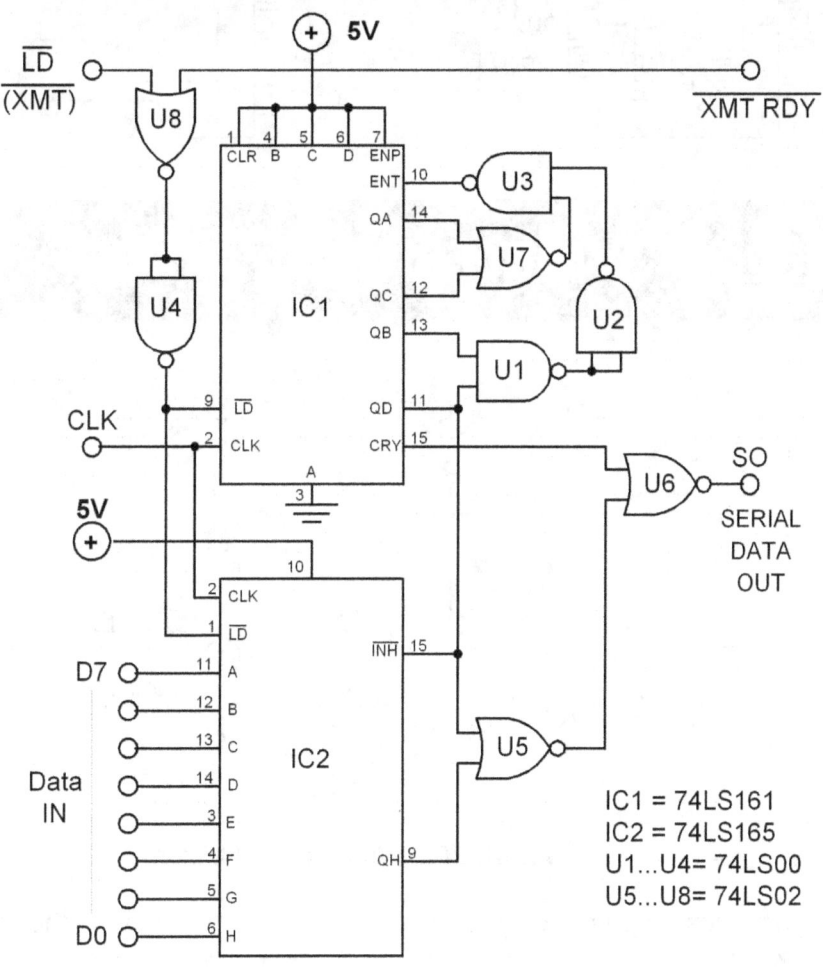

Diagram 69.0 Parallel-Serial Converter

Electronic Circuits 1.2

Parallel to serial converters are used in computer technology for diverse applications. For example it can be used as printer interface for printers with serial inputs or as interface for keyboards with parallel outputs. The circuit is composed of one parallel synchronous binary counter, one synchronous shift register, four NAND gates and four NOR gates. This circuit consumes about 70 mA.

74161 Synchronous Programmable 4-bit Binary Counter

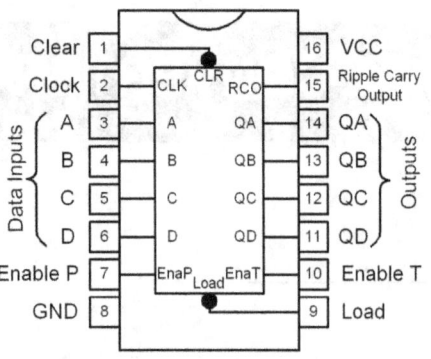

74165 8-Bit Shift Register

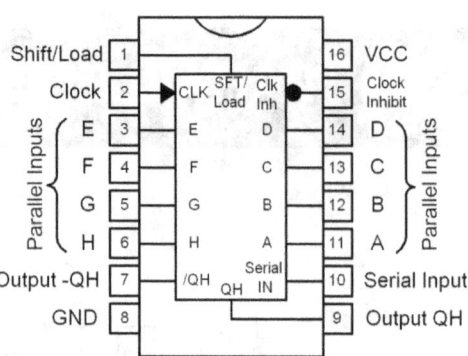

70 VIDEO SIGNAL MIXER

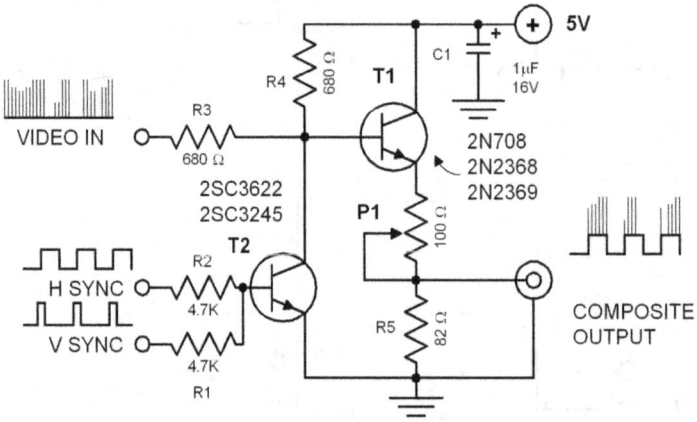

Diagram 70.0 Video Signal Mixer

This mixer combines synchronous and video signals together. It mixes the actual video signal with two control pulses - H sync and Y sync. H and Y sync pulses control the correct reproduction of the video picture in the TV monitor. The mixing of the two sync pulses is accomplished by T2 and all three signals are mixed together by T1. The output amplitude can be adjusted through P1. The mixer circuit can operate up to 25 mHz.

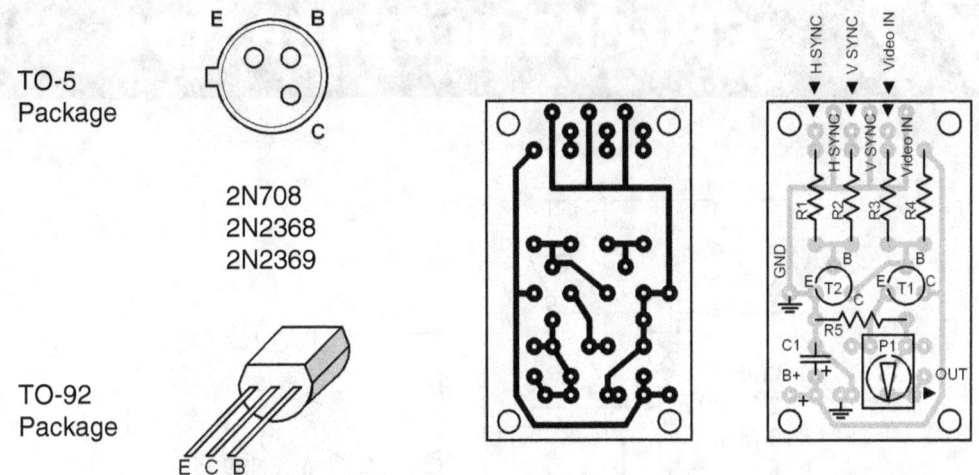

TO-5
Package

2N708
2N2368
2N2369

TO-92
Package

2SC3622
2SC3245

Figure 70.0 Printed Circuit and Parts
Placement Layout for Video Signal Mixer

71 TTL SQUAREWAVE GENERATOR

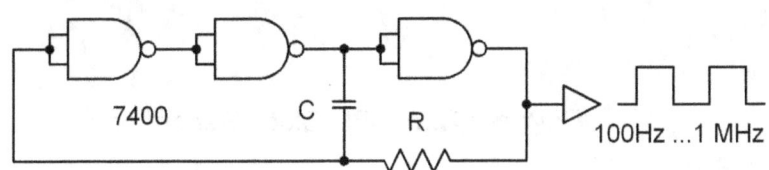

Diagram 71.0 TTL Square Wave Generator

A squarewave signal can be generated from only three TTL gates of a single 7400 TTL logic IC. This circuit can be regarded as a universal oscillator. It functions in a wide range of frequencies and has enough stability for most applications. Its frequency is independent from the supply voltage, starts without any problem and easy to construct. The oscillator frequency is determined by the values of the RC components. The voltage at the inputs of the gate U1 varies between -4 and +6 volts.

The circuit can be redesigned as a variable oscillator by using a 2.2 ohm potentiometer in parallel with R which is the minimal value necessary for the capacitance being used. The circuit can also be constructed from low-power Schottky or CMOS ICs.

72 CPU CLOCK GENERATOR

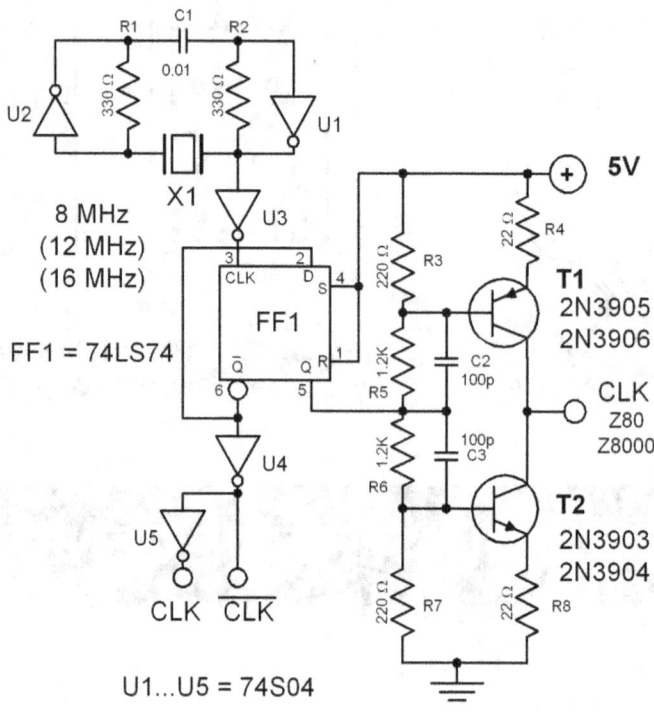

Diagram 72.0 CPU Clock Generator

Generating clock signals for fast CPUs is not always easy. The clock signal must be symmetrical and both high and low levels must be exact. The generator circuit featured here performs excellently in both aspects. The clock circuit works with a quartz oscillator up to 20 MHz.

The oscillator is composed of inverters U1 and U2. The crystal frequency is twice the CPU clock frequency. U3 serves as buffer and the D flip-flop divides the oscillator frequency by 2. The signal from the flip-flop output Q is buffered and inverted (by U3 and U4) and then it is ready to drive other system functions. The output signal is amplified by the driver transistors T1 and T2. The output signal from the driver circuit has the following characteristics.

* high level= 600 mV
* low level= 0.45 V
* rise time= 10 nS if C1= 35 pf

73 I/O SCANNING KEYBOARD

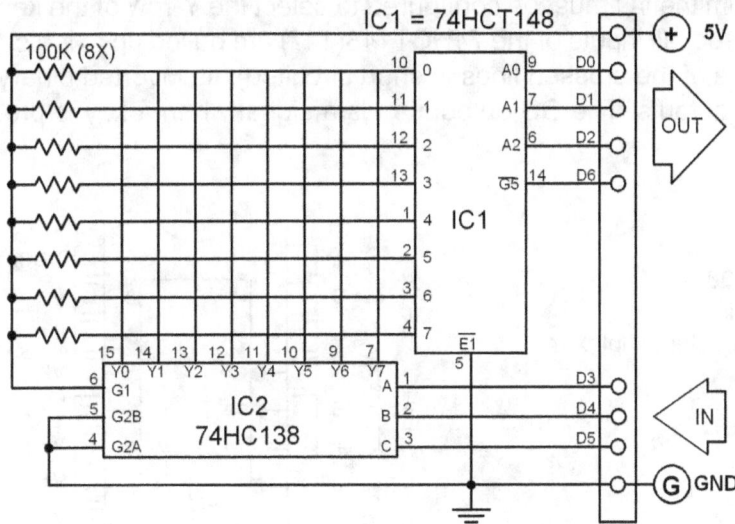

Diagram 73.0 I/O Scanning Keyboard (1)

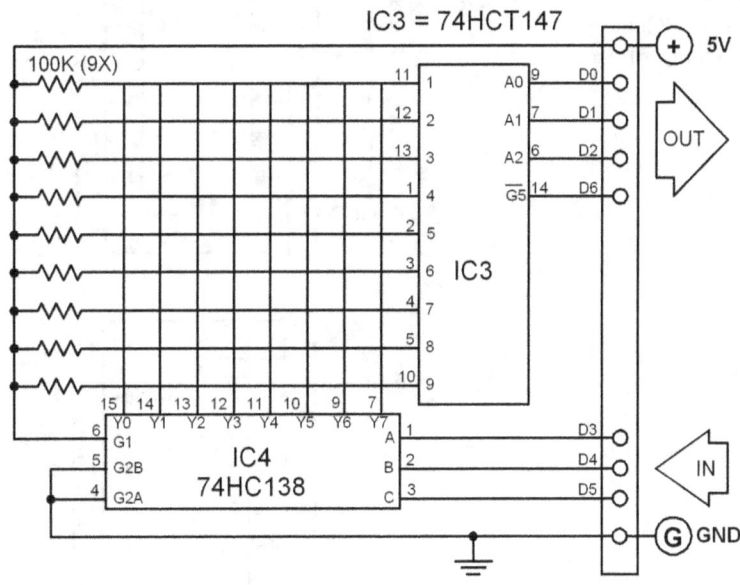

Diagram 73.1 I/O Scanning Keyboard (2)

Electronic Circuits 1.2

Electronic designers and programmers who often work with microcontrollers know the importance of a keyboard that can be connected to I/O lines. The two circuits featured here offer two solutions. The first circuit can decode 56 or 64 keys. The second circuit can decode 72 keys.

Three lines from the I/O must be configured to select the Y-row of the keys. As long as no key is pressed, all inputs of the 74HCT148(147) are pulled up to logic 1. Once a key is pressed (one of the crossed lines is short circuited), its inverted binary form can be read from the outputs. The GS output sends a signal when a key is pressed.

74HC138
3-Bit Binary
Decoder/Demultiplexer

74HCT147
Decimal to BCD
Priority Decoder

74HCT148
Binary 8 to 3
Priority Decoder

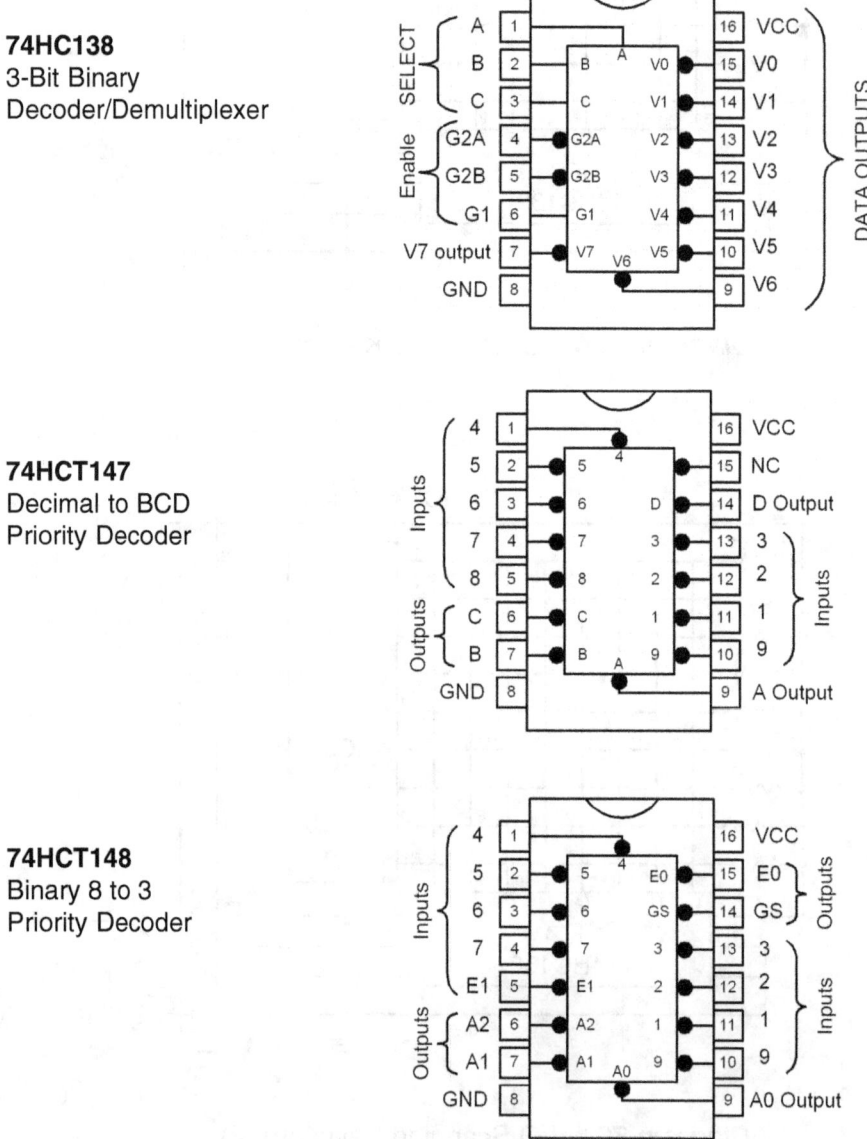

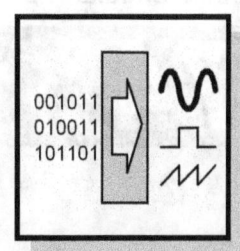

OSCILLATORS & GENERATORS

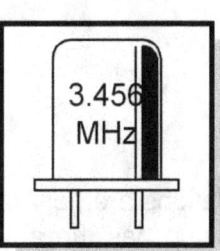

118 Digital Sinewave
119 CMOS Pulse Generator
120 Wien Bridge Oscillator
121 Digital Squarewave
122 Sawtooth Generator
122 48 MHz CMOS Oscillator
123 Power Multivibrator
124 Sawtooth Converter
126 Acoustic Logic Probe

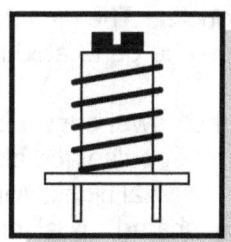

74 DIGITAL SINEWAVE

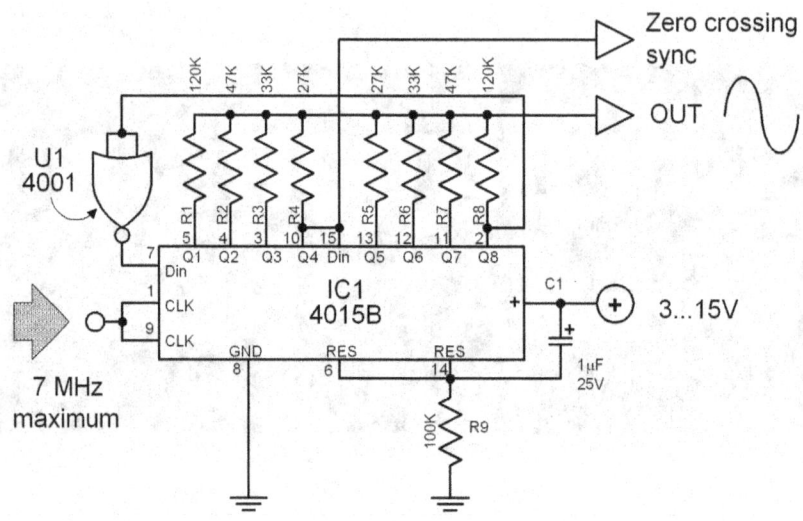

Diagram 74.0 Digital Sinewave Generator

Oscillations are generated today more and more using digital techniques. The digital technique has the advantage that only few components are needed to generate signals with high amplitude constants and variable within a very wide range of frequencies. The circuit shown here generates a sinewave signal. Other signal forms can also be generated by changing the values of R1...R8.

After the power supply is connected to the circuit, the combination R9/C1 generates a short reset-pulse and all outputs of the IC are set to logic 0. The oscillator is controlled by an external clock. With every positive swing of the clock signal, the shift registers of IC1 are shifted 1 position further. After the first input pulse, Q1 becomes 1 and after the 8th pulse, Q8 becomes 1. Once the Q8 becomes 1, the inverter sets the D input to 0. The next series of pulses sets all the registers one by one to 0. After the 7th pulse, Q8 becomes 0 and the process repeats. The 1 and 0 of the registers are converted to a sinewave by the resistors R1...R8. The frequency of the sinewave is 1/8 of the clock fequency. The highest frequency of CMOS ICs is 7 MHz, therefore a sinewave signal of 500 MHz maximum can be generated by this circuit.

A 555 clock generator can be used to drive the register IC. A squarewave signal with the same pulse width and frequency comes out from D_{IN} pin of the IC. It can be used as a trigger signal for an oscilloscope.

75 CMOS PULSE GENERATOR

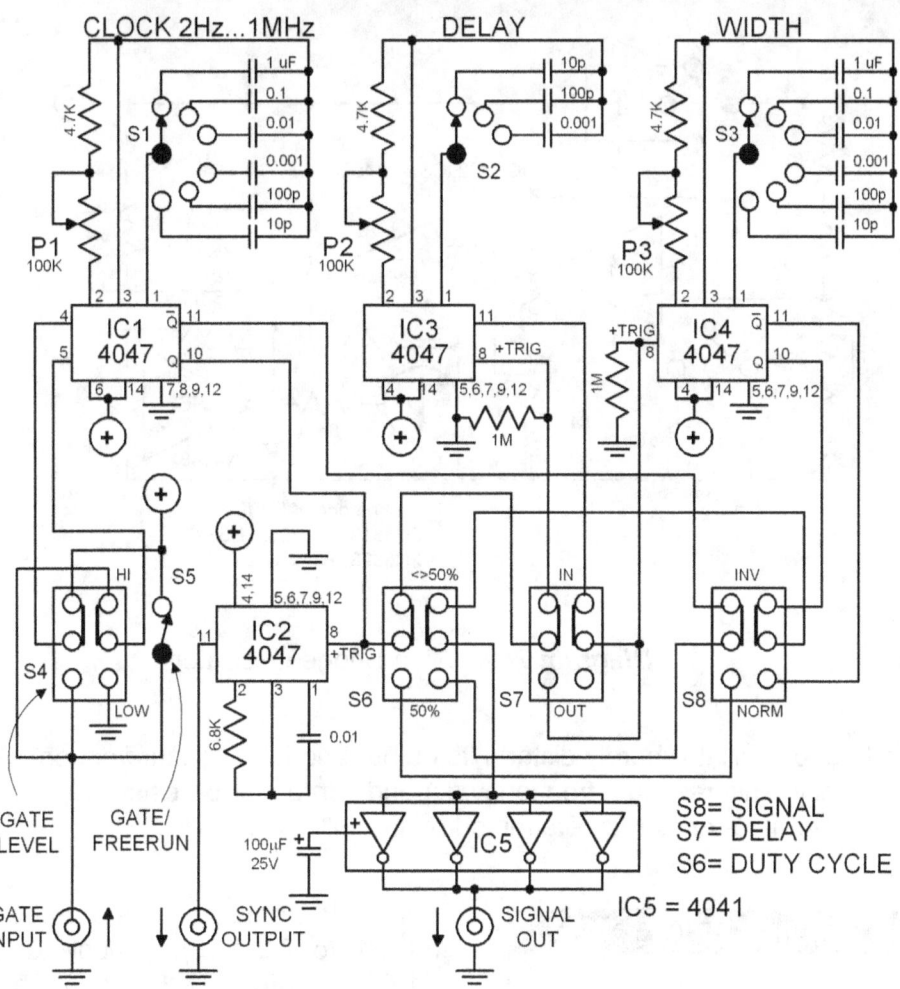

Diagram 75.0 CMOS Pulse Generator

A pulse generator is highly invaluable in developing digital circuits. Such generator must be versatile: the clock frequency must be variable within a wide range of frequencies, the pulse width must be variable and the pulse level must be automatically adaptive.

The generator circuit featured here has all these characteristics. It uses CMOS ICs which give it two advantages. First, the circuit can be powered by batteries. Second, it gives the circuit the capacity to be adaptive to the pulse levels needed by the device under test. The circuit can also be powered from the supply voltage of the device itself, in this way the pulse levels generated by the circuit is automatically adapted to the needed value. The current load is also negligible.

76 WIEN BRIDGE OSCILLATOR

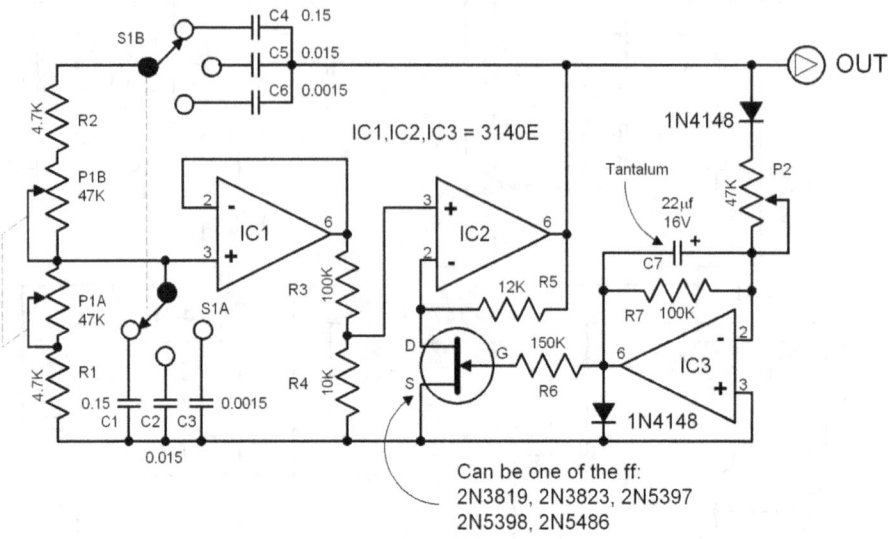

Diagram 76.0 Wien Bridge Oscillator

A Wien bridge oscillator hardly distorts its output and its resonant frequency can be changed easily. This resonant frequency depends on a pair of resistors and a pair of capacitors. The frequency **f** is determined by the following formula:

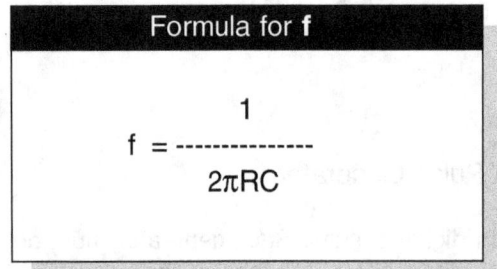

Formula for **f**

$$f = \frac{1}{2\pi RC}$$

Symbolic resistor R is composed of R1+P1a(or R2+P1b). Symbolic C is either C1,C2 or C3(or C4,C5 or C6). The active components of the oscillator are IC1 and IC2. A part of the signal from IC2 is fed to the voltage divider IC3/T1. The FET T1 serves in this circuit as a potentiometer and a part of the feedback of IC2.

The amplification factor of this opamp is voltage dependent and can be changed by changing the supply voltage. Potentiometer P2 controls the stability of the oscillator. The component values in the featured circuit gives an oscillation frequency berween 20kHz and 22.5 kHz with a distortion factor of approximately 2%.

77 DIGITAL SQUAREWAVE

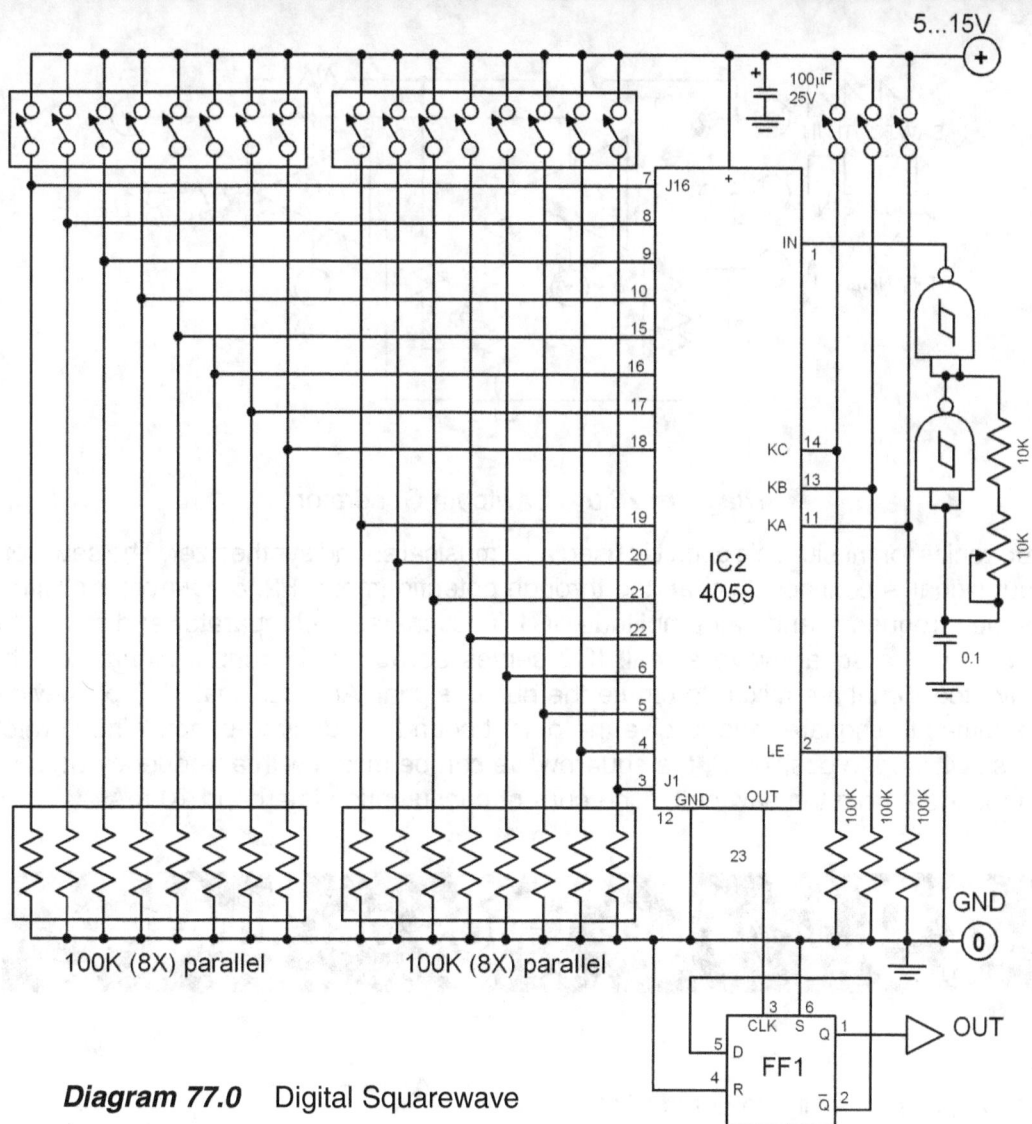

Diagram 77.0 Digital Squarewave

The heart of this circuit is a 4059 divider IC. It can be programmed to divide by a factor of 3 up to 15999 . This division factor is programmed by simply configuring the logic levels at the inputs J1 up to J16 and Ka up to Kc.

The circuit is controlled by an oscillator constructed from a NAND schmitt trigger. A flip-flop is added at the output to convert the signal pulses coming from the divider into a symmetrical form. The flip-flop however further divides the signal by 2:1. Along with the J inputs, the divider has also 3 K inputs. The IC can be programmed by a computer, by an UP-DOWN counter or by manual switching like the one shown in the circuit diagram. The circuit works by a supply voltage betwen 4 and 15 volts. The current consumption is negligible.

78 SAWTOOTH GENERATOR

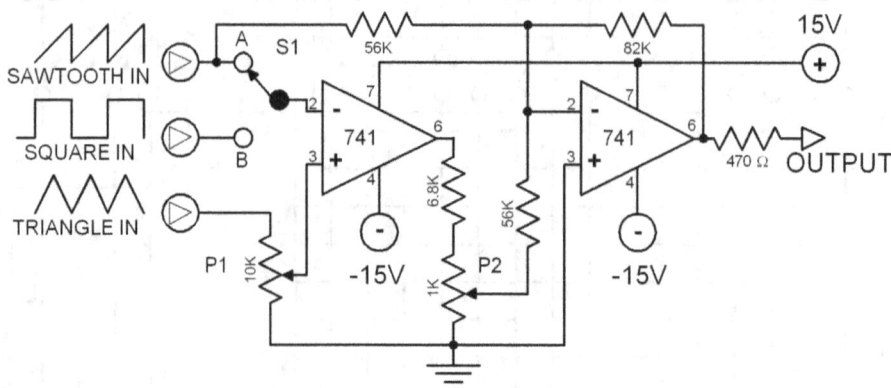

Diagram 78.0 Sawtooth Generator

Electronics for music. This circuit is used as a musical sound synthesizer. The sawtooth input signal is continously changed through potentionmeter P2 to a waveform with a doubled frequency and half amplitude. IC1 functions as a comparator and forms the sawtooth to a squarewave signal. IC2 serves as adder. The input signal and the converted signal are mixed to create the output signal. An additional LFO pulsewidth modulates the squarewave to give the output sound the desired effect. When switch S1 is switched to position B, the squarewave can be mixed with a frequency which is independent from the sawtooth. The current consumption is around 10 mA.

79 48 MHz CMOS OSCILLATOR

Digital gates normally do not oscillate above 30 MHz. The reason lies on the physical limitations of the crystals used which normally do not oscillate properly in such very high basic frequency. This circuit however makes the crystal oscillate to its 3rd overtone. This is done by a LC circuit connected in series to the crystal.

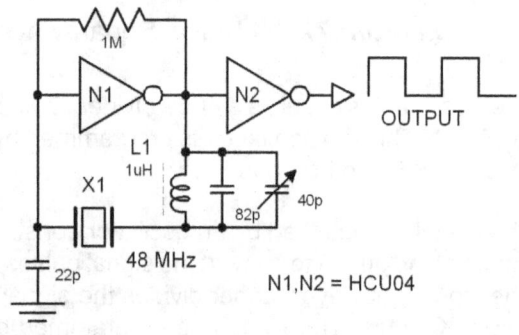

Diagram 79.0 48 MHz CMOS Oscillator

The LC circuit is tuned to 16 MHz so that a crystal with a frequency of 3x16 generates 48 MHz. For digital applications, an inverter must be used to clean the pulses' edges. If non-buffered CMOS ICs are used, the frequency of the circuit can be obtained exactly. HCL family digital gates increase the oscillator frequency up to approx. 60 MHz.

80 POWER MULTIVIBRATOR

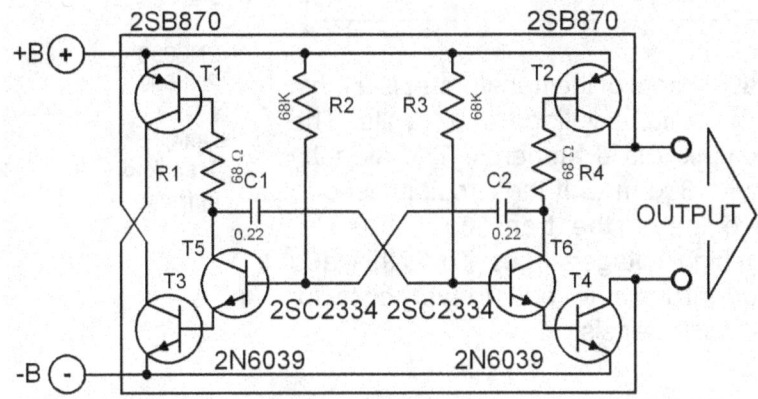

Transistor equivalents:
2SB870 = 2SB921, 2SB826, 2SB633
2SC2334 = 2SD1062, 2SD1237, 2SD1238, 2SD613
2N6039 = 2N6038

Diagram 80.0 Power Multivibrator

This simple multivibrator circuit can drive relatively large loads. It delivers a symmetrical squarewave signal with an amplitude that is dependent on the supply voltage. The astable multivibrator is composed of T5,T6,R1...R4,C1,C2. The collector currents of T5 and T6 switches T1 and T2 while the emitter currents of T3 and T4 drive. The value of R4 determines the current limit and can be fixed according to one's decision. One should take into account that although the transistors can handle large amount of current, their current amplification is small.

The current limitation must then be:

$$hfe(max)(Ub-1.4)/R4$$

If the value for both R1 and R4 is 68 W, the circuit can handle load currents up to 3 amperes. The output frequency of the oscillator can be computed using the formula 0.7(RC). With the component values shown in the circuit and with 12 volts supply, the frequency will be approximately 53 Hz. If the supply voltage is 14 volts, the frequency will be 50 Hz.

Electronic Circuits 1.2

One good application for the circuit is working as a battery-powered voltage converter. In such application, the output of the circuit is connected to a transformer. If it is desired to use the circuit as a voltage converter powering a 40 watt lamp, the following changes in the component values are:

Component changes:		
R1,R4	=	33 ohms
R2,R3	=	33K
C1,C20	=	220µF/parallel
Tr	=	220V/9.5V 5A

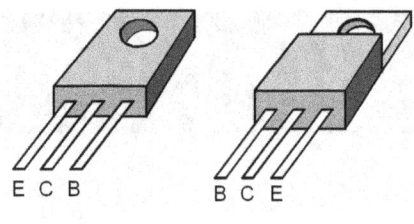

E C B   B C E

2N6038   2SB870
2N6039   2SB921
                2SB826
Note: 2SB633
Transistors   2SC2334
in front view   2SD613
                2SD1062
                2SD1237

The squarewave output of the transformer is almost 240 Vrms with an input voltage of 14 volts. The current consumption is 6 amperes. The standby current is around 300 mA. If the circuit is used to drive inductive loads, the transistors must be protected from high voltage surges. To do this, add two fast highcurrent diodes in series to every collector and emitter of each transistor.

81 SAWTOOTH CONVERTER

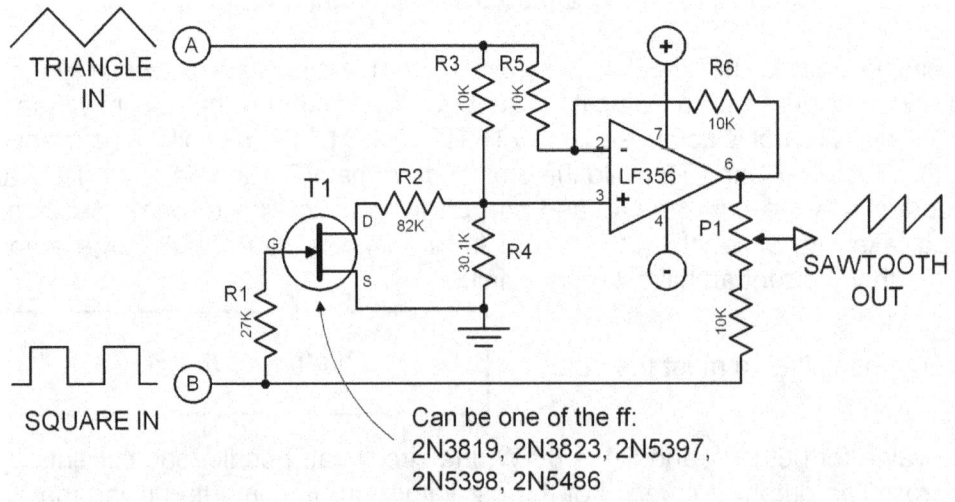

Can be one of the ff:
2N3819, 2N3823, 2N5397,
2N5398, 2N5486

Diagram 81.0   Sawtooth Converter

Most simple function generators generate sinewave, squarewave and trianglewave but a clean sawtooth is rarely available. This circuit forms a sawtooth by combining a squarewave and a trianglewave signal together. The quality of the sawtooth is therefore dependent on several factors: the linearity of the trianglewave, the edges of the squarewave and the phase relation between the trianglewave and the squarewave.

The converter is constructed out of a single opamp which mixes the two input signals. The trianglewave is directly fed to the inverting (minus) input of the opamp while the squarewave is first processed by the FET T1. Out of the opamp output comes a trianglewave with an inverted falling edge. The sawtooth signal available has a doubled frequency. When the DC level of every inverted edge is raised to a certain value so that its lowest level synchronizes with the highest level of the previous edge, a sawtooth with the correct frequency and doubled amplitude can be generated. This quite complicated technique can be easily applied by adding the output signal with the squarewave signal. P1 and P2 are responsible for mixing the two signals. Resistors R2 and R4 are 1% types.

The converter delivers a clean sawtooth signal between 60 Hz and 15 kHz. The supply voltage can be between +/-10 volts and +/-15 volts. The current consumption of each opamp is around 4 to 6 mA.

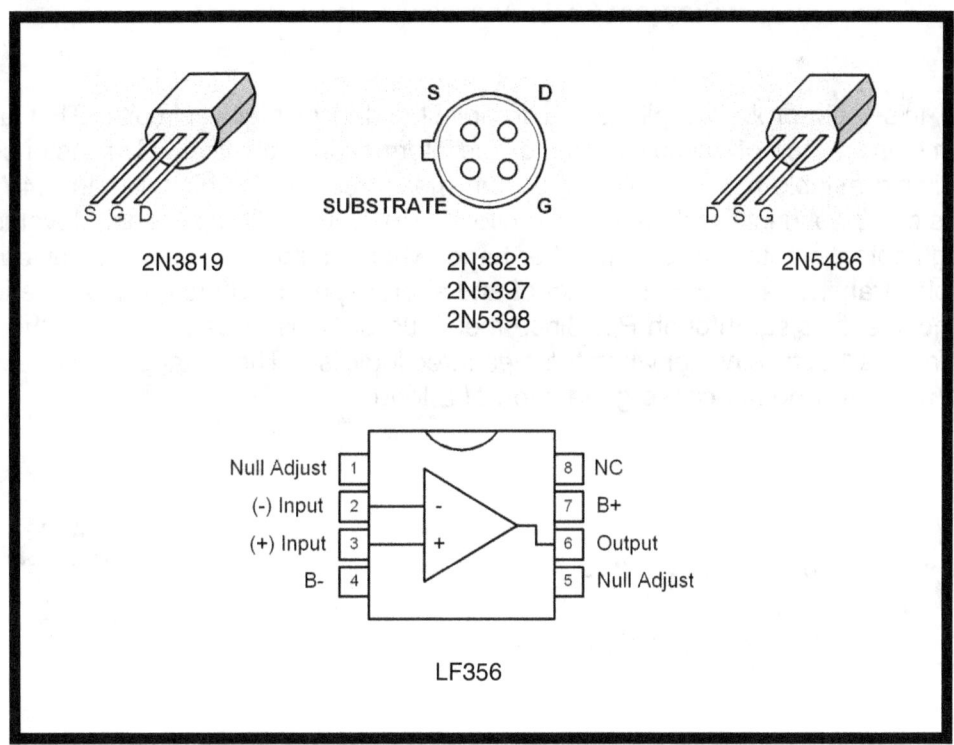

82 ACOUSTIC LOGIC PROBE

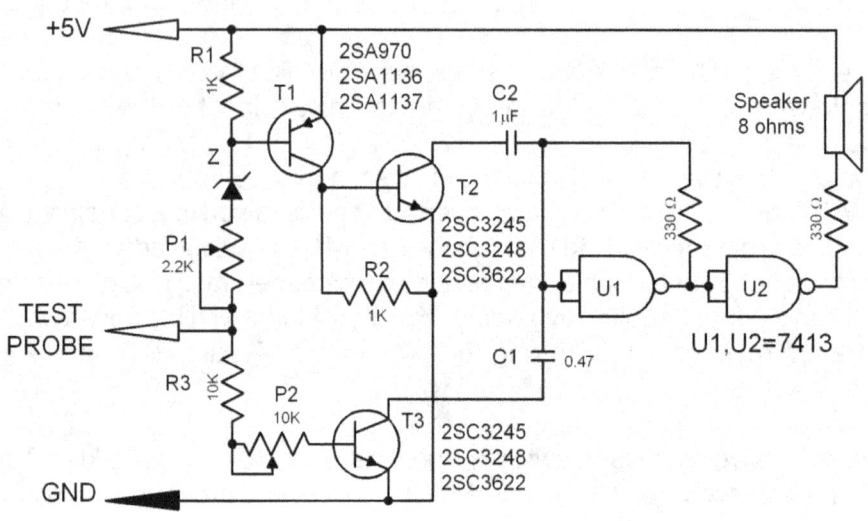

Diagram 82.0 Acoustic Logic Probe

Using this logic probe simplifies the troubleshooting of digital circuits. This probe determines the logic of a circuit and sends the information acoustically. The tester circuit is a simple astable multivibrator. When the measured logic is higher than 2.4V, the transistor T3 conducts and shorts capacitor C1 to ground. This threshold level is set through potentiometer P2. On the other hand, when the measured logic is lower than 0.8 volts, transistors T1 and T2 conduct and short capacitor C2 to ground. The lower threshold level is set through P1. Since the value of C2 is twice that of C1, the tone generated is 1 octave higher when the measured logic is 1. The voltage levels between 0.8 V and 2.4 V do not cause generation of a tone.

The supply voltage +5V is taken from the supply line of the circuit being tested.

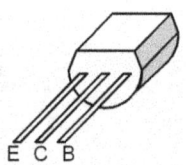

2SA970
2SA1136
2SA1137
2SC3245
2SC3248
2SC3622

E C B

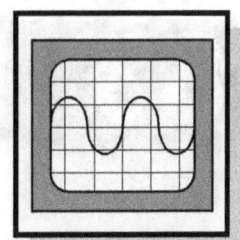

TESTERS & MULTIMETERS

128 FET Voltmeter

129 Lamp & Fuse Tester

130 Continuity Tester

131 Audio Signal Injector

132 Analog Frequency Meter

133 Transistor Tester

134 FM Transmitter

135 Servo Tester

136 Digital Sample & Hold

137 Infrared Detector

138 Mega-W Multimeter

139 Car Thermostat

140 Opamp IC Tester

141 Mini Frequency Meter

142 Trigger Amplifier

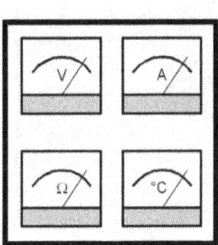

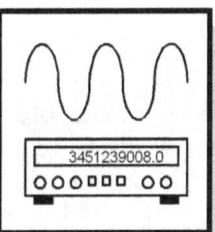

83 FET VOLTMETER

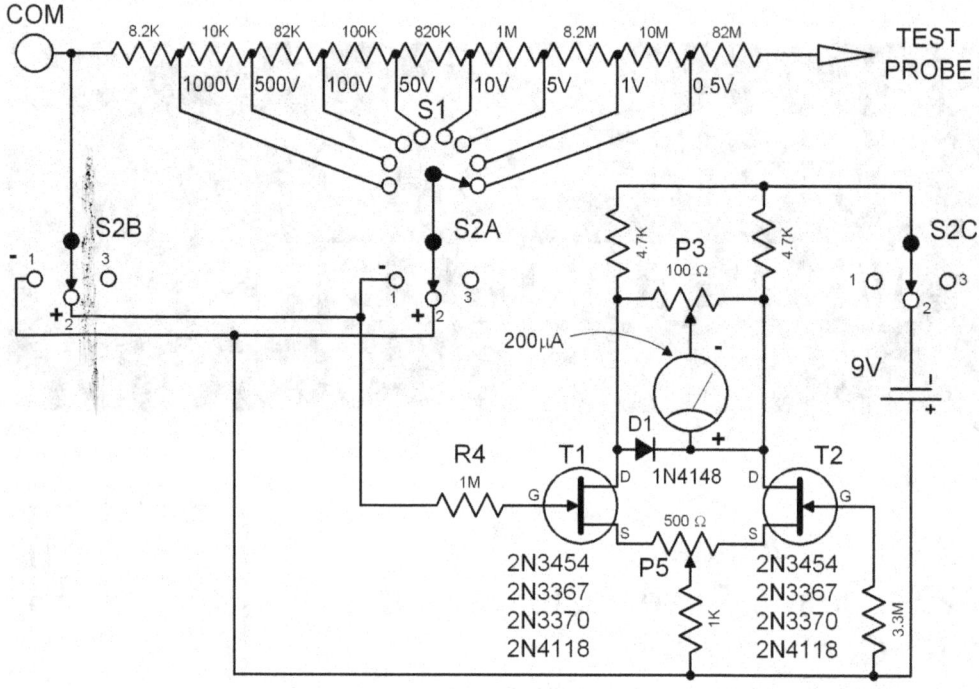

Diagram 83.0 FET Voltmeter

Field effect transitors -commonly called FET's- have very high impedance input characteristics that make them ideal for a voltmeter circuit. The high impedance reduces the loading effect of a voltmeter to the actual circuit under test. This makes the test results highly accurate.

In the featured voltmeter circuit, both FETs T1 and T2 build a differential amplifier which gives an output of 500 mV for a full range meter deflection. The zero voltage calibration is done through potentiometer P5. The end of scale calibration is done through P3. Diode D1 serves as overload protection by limiting the input voltage to a maximum of 700 mV. The polarity of the testprobes can interchanged through S2.

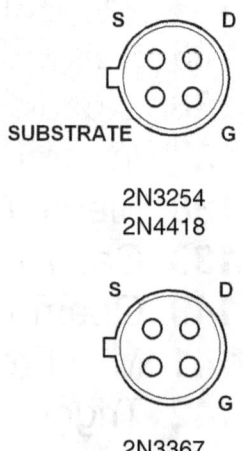

SUBSTRATE

2N3254
2N4418

2N3367
2N3370

84 LAMP & FUSE TESTER

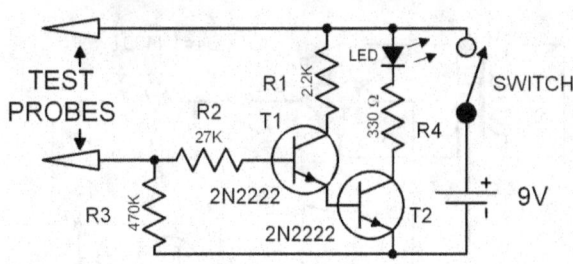

Diagram 84.0 Lamp & Fuse Tester

Testing cables, wires, fuses, lamps, etc. belongs to a repair job like butter to a bread. Sometimes, this becomes too cumbersome since one has only two hands and too often, one has to hold the part being tester and the two probes of an ordinary continuity tester all at the same time. It wouldn't be much of a problem if only one had an assistant. A third hand? Well, you never have seen a mutant repairman yet, have you?

This circuit enables easy testing of lamps and fuses by using the conductivity of the human body. One of the test probes is connected to the part under test while the other probe is held by the normal hand (bare, of course). When the lamp or fuse is working properly, a small amount of curent flows through the hand which is enough to switch on the transistors and light the LED.

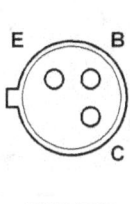

2N2222

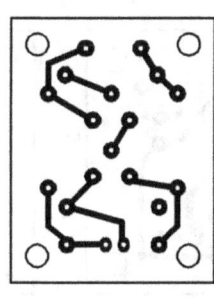

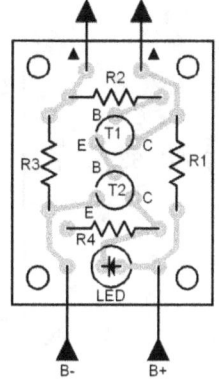

Figure 84.0
Printed Circuit Layout
for the Lamp & Fuse Tester

Figure 84.1
Parts Placement Layout
for the Lamp & Fuse Tester

85 CONTINUITY TESTER

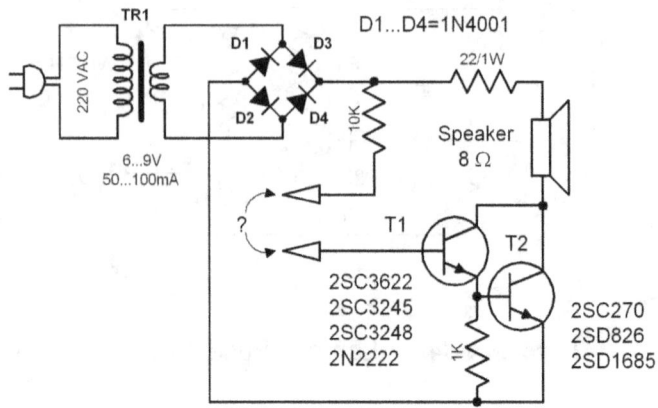

Diagram 85.0 Continuity Tester

Technicians often need to test large volume of electrical wirings or cables. A multimeter would be inconvenient in this case since one would have to hold the cables and tester probes and at the same time read the meter scale.

The easiest way is to use a tester that produces a tone when a cable is not broken. The circuit featured here produces a 60 Hz tone that is generated by the pulsating DC coming from the bridge rectifier.

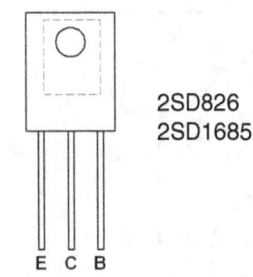

2SD826
2SD1685

E C B

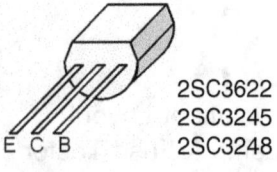

E C B

2SC3622
2SC3245
2SC3248

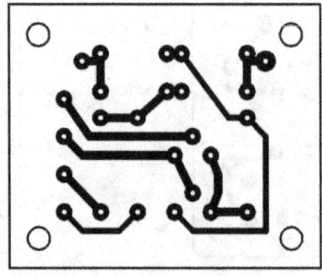

Figure 85.0
Printed Circuit Layout

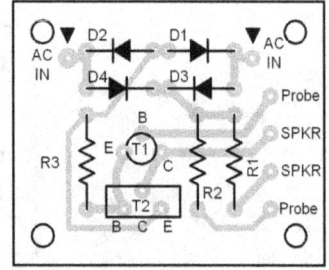

Figure 85.1
Parts Placement

86 AUDIO SIGNAL INJECTOR

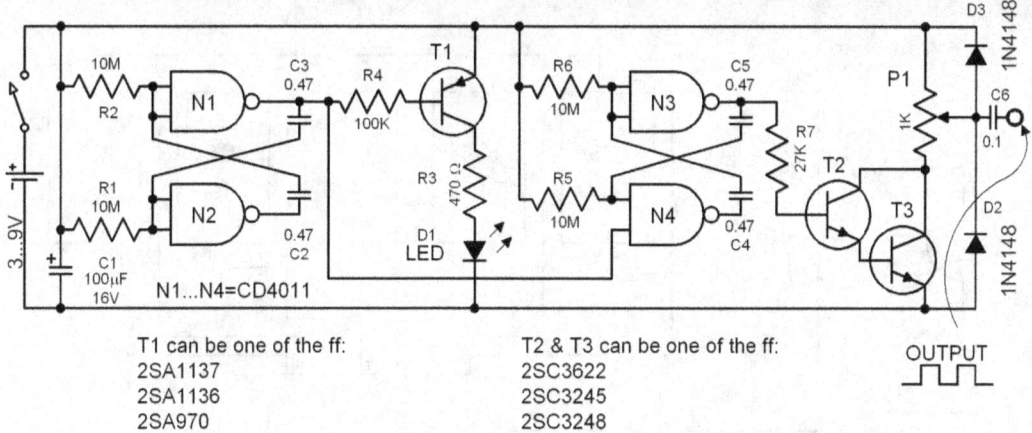

T1 can be one of the ff:
2SA1137
2SA1136
2SA970

T2 & T3 can be one of the ff:
2SC3622
2SC3245
2SC3248

OUTPUT

Diagram 86.0 Audio Signal Injector

In troubleshooting audio amplifiers, a signal injector helps to simplify and speed up pinpointing of the trouble's cause. Common signal injectors generate a frequency of 1 kHz but is was found out in practice that the capability to pinpoint the troublesome component will be improved if the signal being injected is also keyed.

The circuit featured here offers this keyed signal characteristic. It is made of two astable multivibrators with different oscillation frequencies. The gates U3 and U4 oscillate at 1 kHz. This is the actual audible signal injected into the circuit being tested. This frequency is determined by the values of R5,R6, C4 and C5. This 1 kHz signal is keyed - meaning it is periodically switched on and off - by the gates U1 and U2. The keying frequency is determined by R1,R2,C2 and C3. LED3 gives an optical signal of the keying. The output signal level can be varied through potentiometer P1.

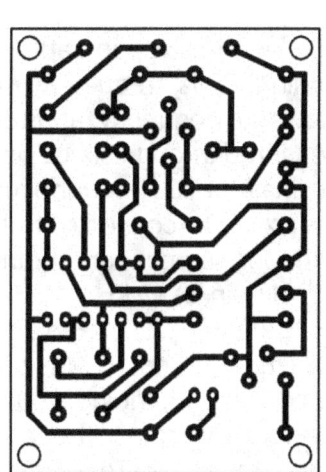

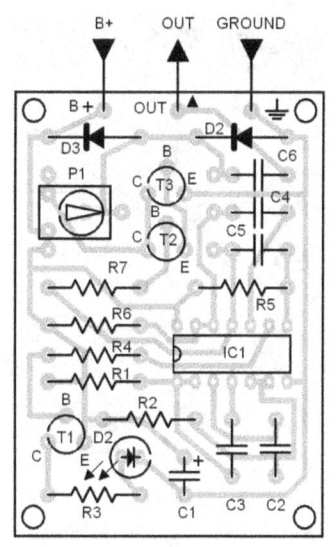

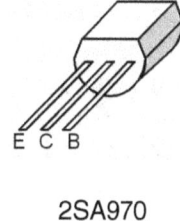

2SA970
2SA1136
2SA1137
2SC3622
2SC3245
2SC3248

Figure 86.0 Printed Circuit Layout   *Figure 86.1* Parts Placement Layout

87 ANALOG FREQUENCY METER

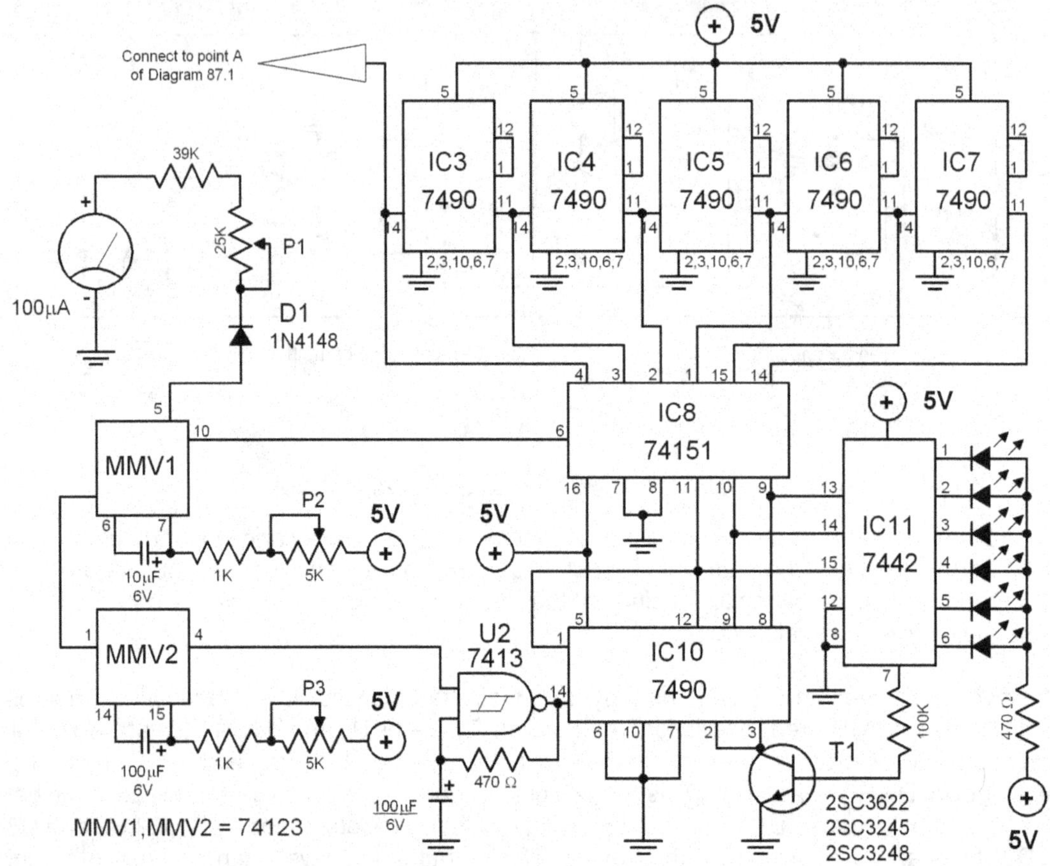

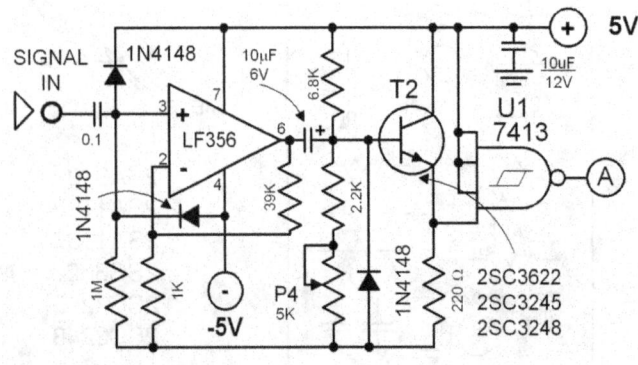

Diagram 87.0 Analog Frequency Meter

Diagram 87.1 Analog Frequency Meter
(input module)

Simple, effective, and automatic ranging! This frequency meter shows the frequency range being measured through six LEDs and the analog value through a meter. The six LEDs (connected to IC11) stand for six measuring ranges: 100 Hz(pin 6), 1 kHz(pin 5), 10 kHZ(pin 4), 100 kHz(pin 3), 1MHz(pin 2) and 10 Mhz(pin 1). Since it is autoranging, you don't need to switch to the correct range. In fact, there is no range selector in this circuit.

To calibrate the circuit, set potentiometers P3 and P4 at midposition. Set potentiometer P1 at maximum and set pot P2 at minimum. Inject a 100 Hz signal with an amplitude of at least 1 volt into the input of the circuit. While this signal is at the input, adjust pot P3 until the LEDs light one after the other. This shows that the multiplexer is working. After that, adjust pot P2 until only the LED for 100 Hz (IC11 pin 6) remains lighted. While this LED is on, slowly adjust potentiometer P1 until the analog meter settles to its full meter deflection. Finally the sensitivity of the input is optimized through P4.

88 TRANSISTOR TESTER

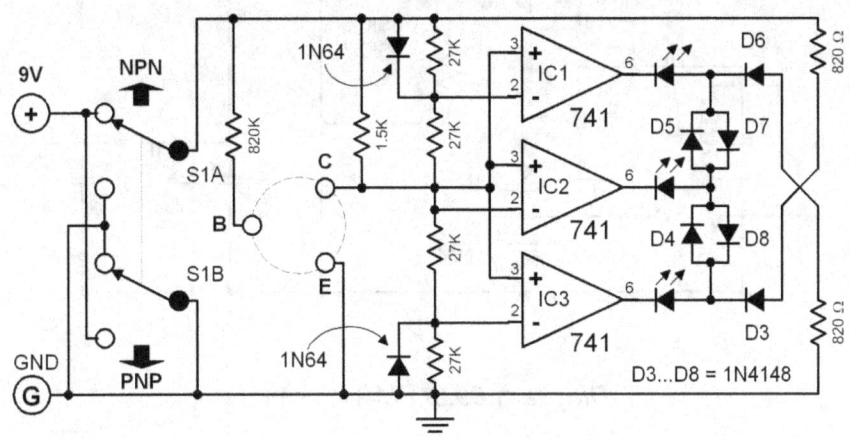

Diagram 88.0   Transistor Tester

Speed up your troubleshooting work. Use this highly reliable transistor tester. It is very handy and shortens checkup time since you don't need to desolder the transistor out of the circuit board to test them. The tester checks them up right in the circuit board. How is it possible you may ask. Well, simple. This circuit tests the amplification factor of small transistors and shows it through three LEDs. The amplification ranges shown by the LEDs are:

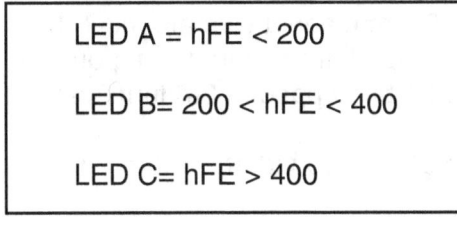

LED A = hFE < 200

LED B= 200 < hFE < 400

LED C= hFE > 400

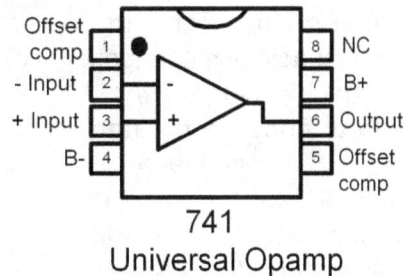

741
Universal Opamp

During the test, the collector current of the transistor under test is approximately 2mA. The circuit can be used to test either an NPN or PNP transistor.

89 FM TRANSMITTER

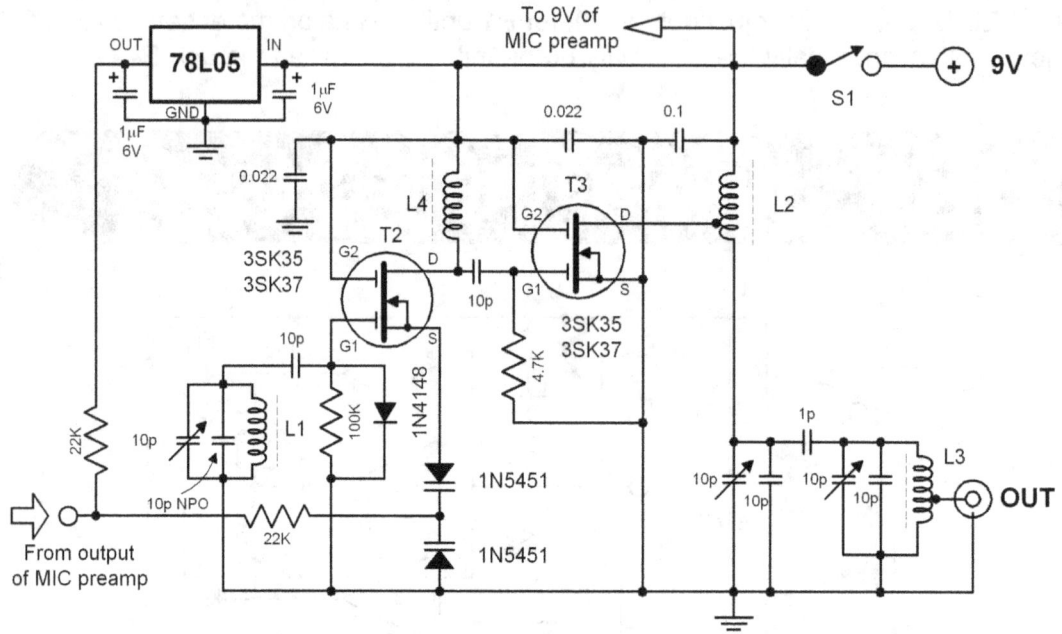

Diagram 89.0 FM Transmitter

This transmitter circuit is designed to be used for aligning FM radio receivers. It has a microphone amplifier with an automatic gain control. The FM modulation is a voltage oscillator and the final stage is built with MOSFETS. A lowpass filter at the output assures the transmission of a clean 10.. 50 mW FM signal. IC1 works as an amplifier supplying a negative control voltage to FET T1. The amplitude of the amplified audio signal can be adjusted through P1. The carrier frequency is generated by the oscillator T2.

The oscillation is frequency modulated through the internal capacitance changes of the double diode D4/D5. The frequency modulated signal appears at the drain terminal of T2. This signal is then amplified by T3 and passes through the lowpass filter before finally arriving at the antenna.

Coil Data
L1= 6 turns 0.8 mm magnet wire around T50-12 ferrite ring, tapped at 2nd turn from ground
L2= same as L1 but tapped at 3rd turn from ground
L3= same as L2 but tapped at 1st turn from ground
L4= 4 turns 0.8 mm magnet wire in a ferrite core.

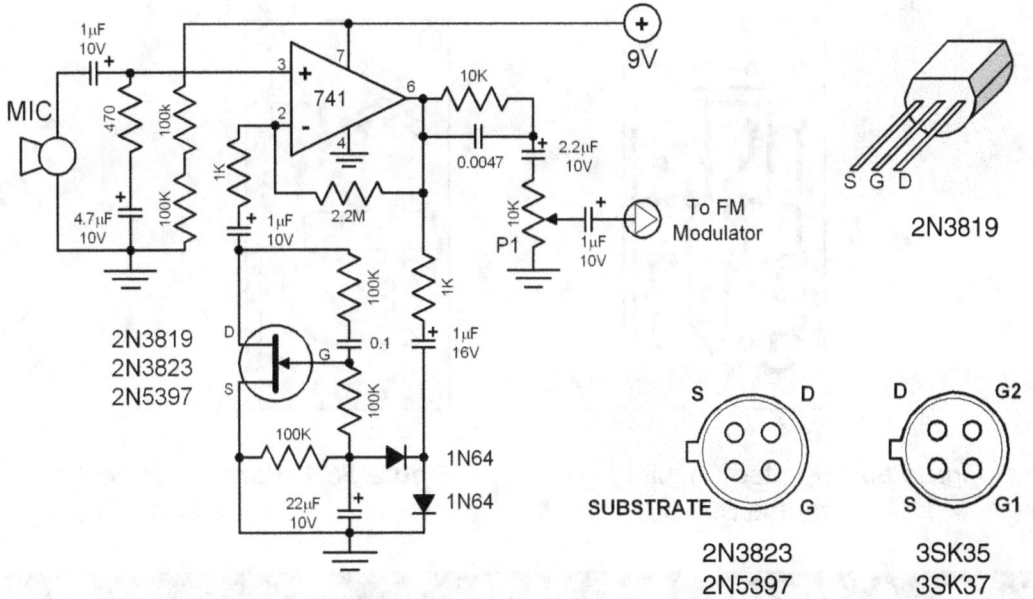

Diagram 89.1 Microphone Preamp
for the FM Transmitter

90   SERVO TESTER

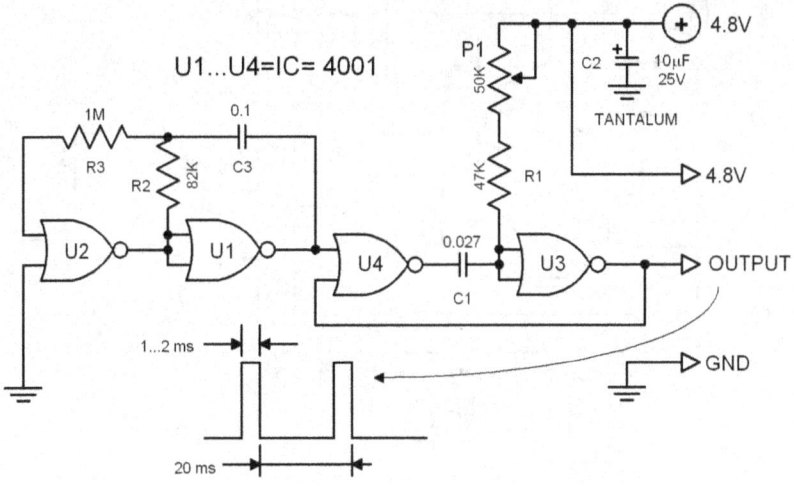

Diagram 90.0   Servo Tester

This small circuit is used to test a servo without using the main control transmitter. It generates a pulsewidth modulated signal. The pulsewidth of the signal is 1.5 mS at the neutral position. The pulsewidth is 1 mS for one end position and 2 mS for the other end. The pulse width is varied through P1 and swings between 1 mS and 2 mS.

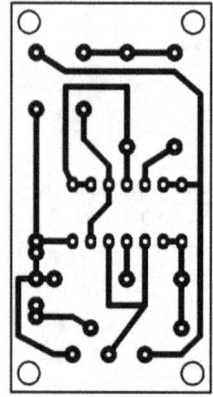

Figure 86.0 Printed Circuit Layout
for the Servo Tester

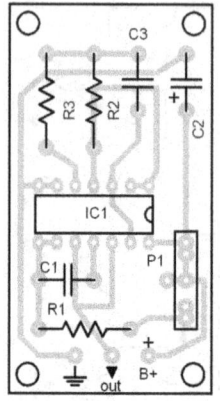

Figure 86.1 Parts Placement
for the Servo Tester

91 DIGITAL SAMPLE & HOLD

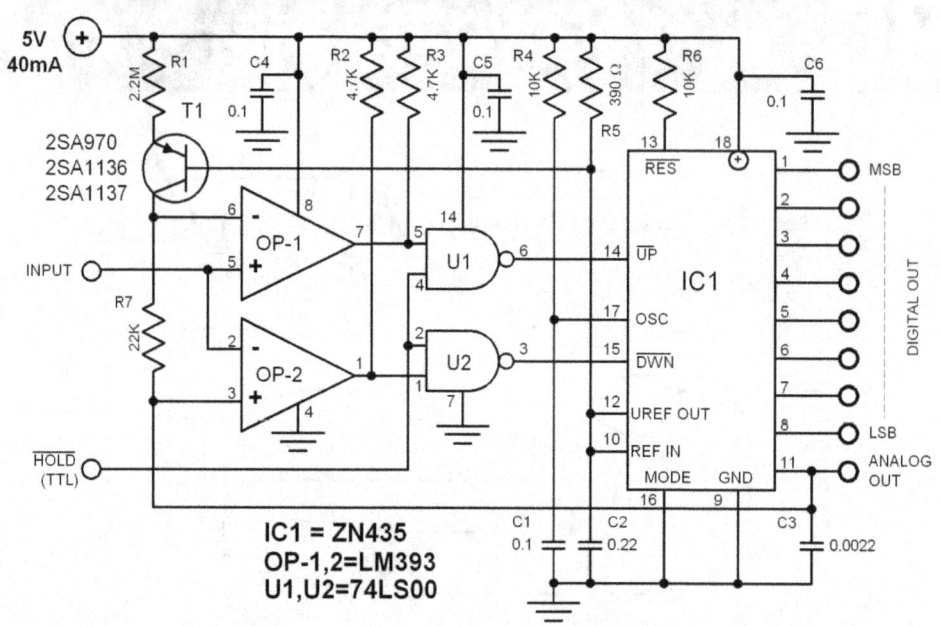

Diagram 91.0 Digital Sample-&-Hold

Since sample-and-hold circuits are used as temporary storage for analog signals, these circuits are usually constructed with analog components. However, analog circuits have a big disadvantage - drift. That is why in such applications, a digital circuit works much better.

The digital sample-and-hold circuit featured here temporarily stores an analog signal in a digital way! How? First, the analog signal is converted to a digital one by an A/D converter and then reconverted back to analog value by a D/A converter. To save money and to simplify the construction of the circuit, only an A/D converter and an up-down counter are used. This combination prevents the use of a second converter. To achieve the desired result, the input voltage is continously compared to the output voltage of the A/D converter through a window comparator. Once the input voltage changes, the comparator triggers the counter to count either upwards or downwards depending on the direction of the input voltage's change. The counter continues to count until the output voltage and the input voltage are again equal. The result of the A/D conversion can then be taken from the output terminals of the counter. From this "pre-processing" phase, the analog value can be „sampled"and „held" easily. Once the counter is blocked by an Enable pulse, the value of the counter is held for the moment. Since the counter cannot count further, the analog value can be conserved.

The converter is an 8-bit-ADC/DAC ZN435. The internal voltage reference of 2.55 V and the oscillator are set by R5/C2 and R4/C1 respectively so that the oscillator will function to its maximum frequency of 400 kHz. The counter has three control inputs: up, down and mode. The logic level at mode input determines whether the counter stops or counts further after it reaches its maximal count of 255. The counter here is wired to further count after 255. The gates U1 and U2 block the counter. The threshold voltage at IC1 is set to a higher level of 20 mV than OP. This 20 mV creates a dead zone which is necessary to prevent the counter's LSB from oscillating and to suppress unwanted influence of the input's offset voltage to the comparator. The maximum conversion time is 60 nS. The output impedance is quite high - 4KΩ.

92 INFRARED DETECTOR

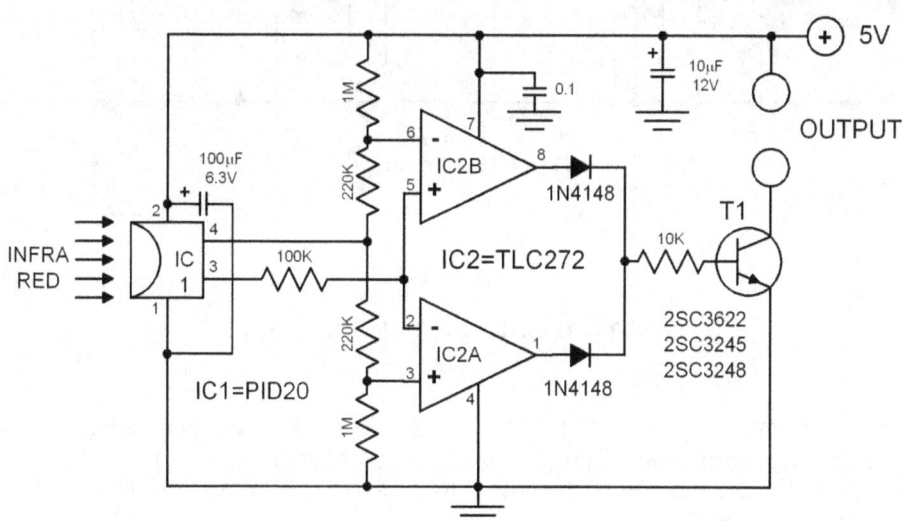

Diagram 92.0 Infrared Detector

Electronic Circuits 1.2

This detector uses a passive IR-detector PID20. It converts heat radiation into electrical impulses. The output voltage of the PID20 increases when an object comes close to it. This object must be warmer than the surroundings. Conversely, its output voltage decreases when the warm object goes away or when an object which is colder than the surrounding comes close to it.

The voltage changes at the output of the IR sensor are sensed by the comparators IC2a and IC2b. Once the sensor detects an object (independent of the object's actual temperature), one of the comparators will turn on the transistor T1. This transistor then closes the relay connected to it. A separate transistor for each comparator can be used so that the sensor will be able to determine whether the object is warmer or colder than the surroundings. The circuit consumes around 1 mA and the sensor consumes 0.2 mA.

93   MEGA-Ω MULTIMETER

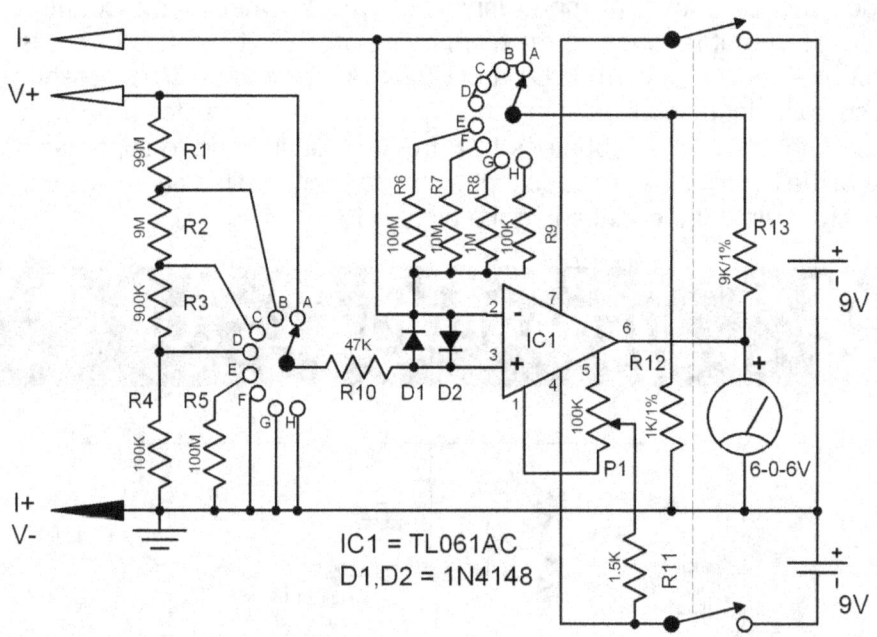

Diagram 93.0   Mega-Ω Multimeter

This DC multimeter offers the advantages found in a high input impedance tester. It measures DC voltage and current. An opamp TL061 is used in this circuit to measure the DC voltage. It is configured to amplify the input level 10 times. In measuring currents, the DC current flows through resistors R6 up to R9.

Since all these resistors have higher values than resistor R12, the output voltage is 10 times higher than the voltage drop at the connected resistor. This way, the tester is able to measure very low current levels. The current ranges are: 6 nA, 60 nA, 600 nA and 6 µA.

Potentiometer P1 functions as a zero ohm adjuster. Since the supply voltage is symmetrical, both voltage polarities can be measured without interchanging the tester probes.

Measurement Ranges
A= (-)0.6 - 0 - (+)6V
B= (-)6   - 0 - (+)6V
C= (-)60 - 0 - (+)60V
D= (-)600 - 0 - (+)600V
E= (-)6   - 0 - (+)6 nA
F= (-)60 - 0 - (+)60 nA
G= (-)600 - 0 - (+)600 nA
H= (-)6   - 0 - (+)6 µA

94 CAR THERMOSTAT

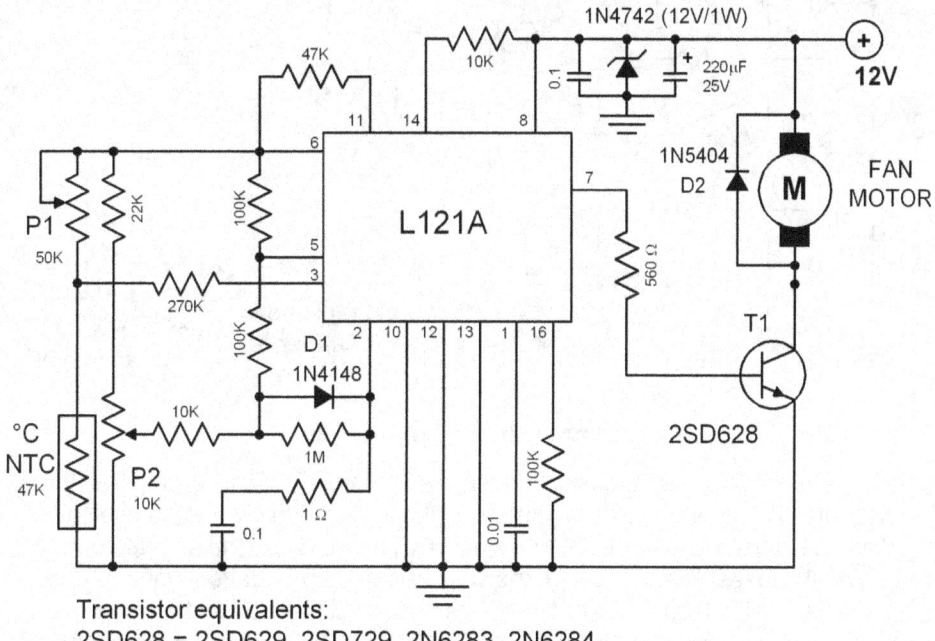

Transistor equivalents:
2SD628 = 2SD629, 2SD729, 2N6283, 2N6284

Diagram 94.0 Car Thermostat

This circuit regulates the air blower system of a car. When the temperature inside the car exceeds a certain preset level, transistor T1 turns on the air blower. A temperature dependent resistor NTC is used as a sensor. The resistance of this resistor must be 47 kilohm at 25°C. The circuit is calibrated through potentiometers P1 and P2 so that the air blower will turn on once the threshold temperature is reached. P2 sets the "ON" level and P1 sets the "OFF" level. To calibrate the circuit: first set P1 at middle position and slowly adjust P2 until the air blower turns on. The temperature value must be the desired threshold level at this moment. Later when the temperature has gone down, adjust P1 until the air blower stops.

95   OPAMP IC TESTER

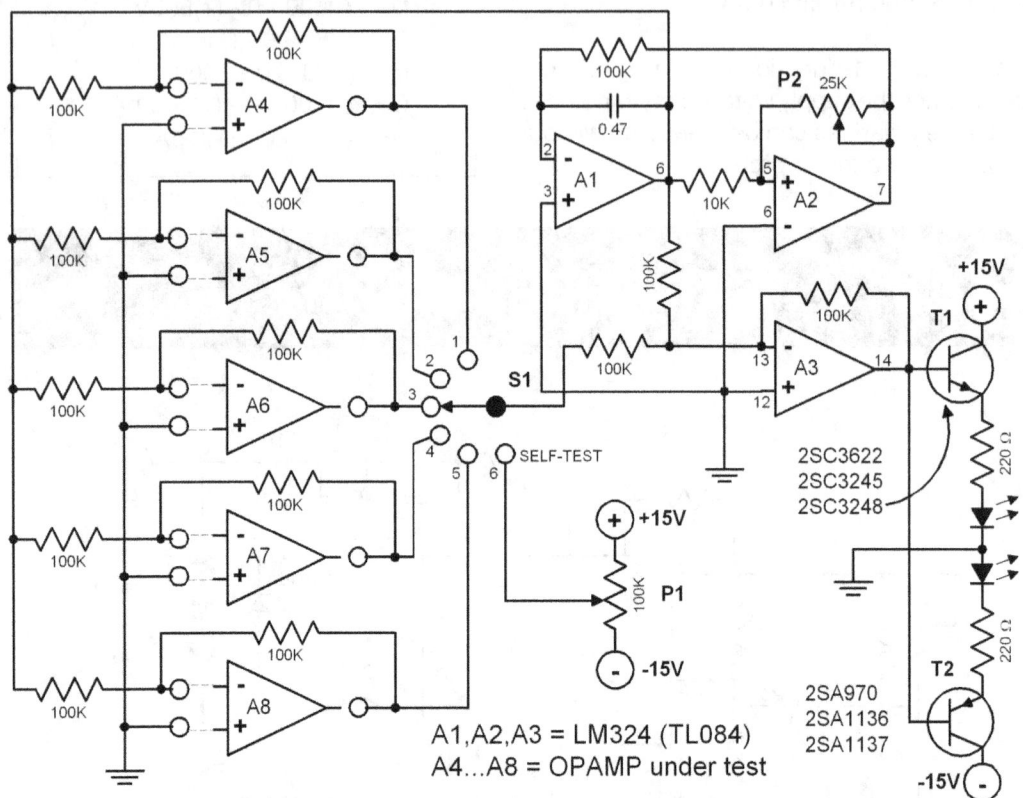

Diagram 95.0 OPAMP IC Tester

Any type of opamp IC can be tested by using this tester circuit. Testing is simple. A trianglewave signal is fed into the inverting (minus) input of the opamp-under-test. Since the opamp is configured as inverting amplifier, its output signal is inverted. When the inverted output is added to the input signal, the result is zero volt. Otherwise, the opamp is malfunctioning. The tester has also a self-test capability. Opamps IC1 and IC2 in the circuit function as trianglewave generator and supplies the input signal to the opamps-under-test. IC3 is the summing amplifier. The opamps-under-test are IC4 ...IC8. The transistors drive the LEDs.

How does one determine that the opamps-under-test are not defective? When the opamps are tested one by one (selected by S1) and the LEDs do not blink, then the opamp IC is functioning normally. On the other hand, when the LEDs blink the Opamp IC is malfunctioning.

To calibrate the circuit, one must insert at the test socket a known-to-be-intact opamp. Then P2 must be adjusted until the LEDs are almost about to blink (not yet blinking). The self-test is done in the following manner: While P1 is turned from one end to the other end, a LED will blink at first and then followed by the blinking of both LEDs, finally only one LED will remain blinking. Position no. 6 of S1 turns on the self-test function.

96 MINI FREQUENCY METER

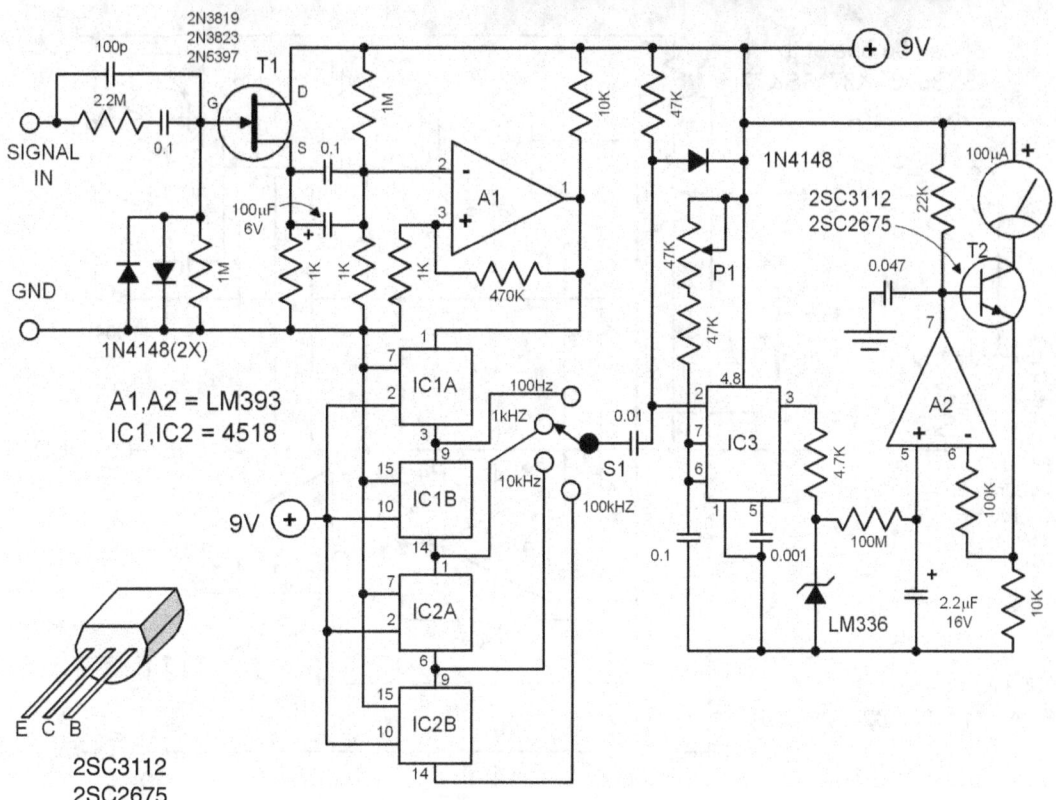

Diagram 96.0 Mini Frequency Meter

This frequency meter is simple to build and can be constructed in a portable format. It can measure frequencies with a minimum input level of 10 mVrms and a maximum frequency of up to 100 kHz. It can be powered by penlight batteries since its standby current is only around 4 mA. It is also very easy to calibrate. The input is protected from high voltage levels up to 250 VAC.

FET transistor T1 serves as buffer for the schmitt-trigger IC1. This opamp is wired to give a hysteresis of only 18 mV so that its sensitivity is enough for the whole frequency range. The output of IC1 is directly connected to the counter IC1a. This counter divides the signal by 2 and the following three counters divide by 10. In any of the measurement range, 50 Hz is the point where the meter will have full deflection. The signal from S2 is used to trigger the monostable multivibrator. The monotime of this circuit is set to one-half of the frequency range - 10 mS. The circuit is actually set to 8 mS.

The pulse/pause relationship of the output signal from IC3 is directly proportional to the input frequency. This pulse is converted to 2.5 Vp-p and integrated to a DC voltage. The circuit IC2 and T2 is a simple voltage to current converter.

The tester is very sensitive that the power line frequency will be displayed when the input is touched by a finger. The circuit can be calibrated using the power line frequency. While touching the input with a finger, adjust P1 slowly until the correct frequency is displayed.

97 TRIGGER AMPLIFIER

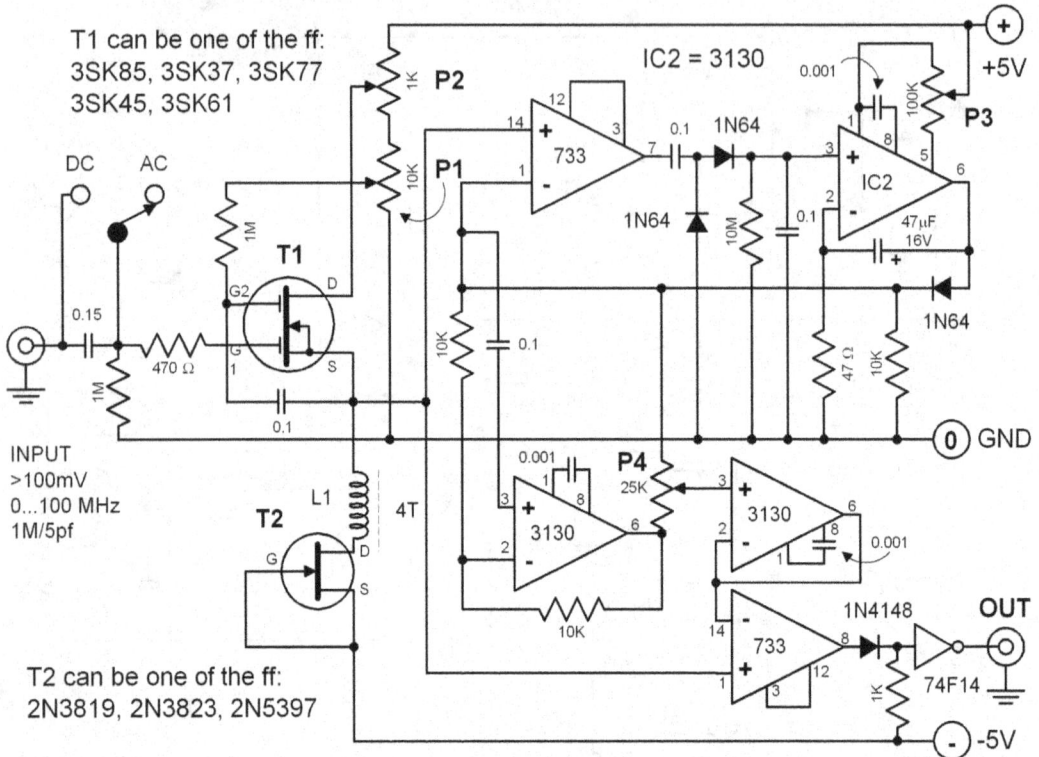

Diagram 97.0 Trigger Amplifier

This trigger amplifier circuit is peak independent and automatically adjusts itself to the signal level. This makes the triggering of oscilloscopes and frequency counters much easier. The circuit is fast enough to process an input signal up to 100 MHz with a sensitivity of 100 mVp-p.

Potentiometers P1 and P2 determine the potential at the source terminal of T1. The trigger sensitivity can be adjusted through P4. Coil L1 is 4 turns (0.22 mm wire diameter) magnet wire in a ferrite core. To obtain maximum sensitivity, short the input terminals and let the circuit operate for several minutes. After several minutes of this idle-run, adjust P1, P2 and P3 carefully until the offset is minimal.

Do not forget to remove the short at the input after calibration.

AUXILIARY

144 Digital Bike Tacho
145 Auto Car Warning
146 Diode Thermometer
148 Battery Monitor
149 Car Antenna Amplifier

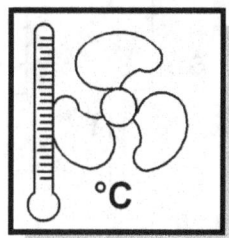

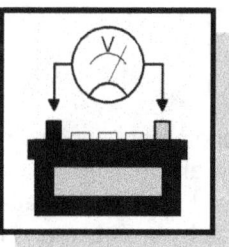

98 DIGITAL BIKE TACHO

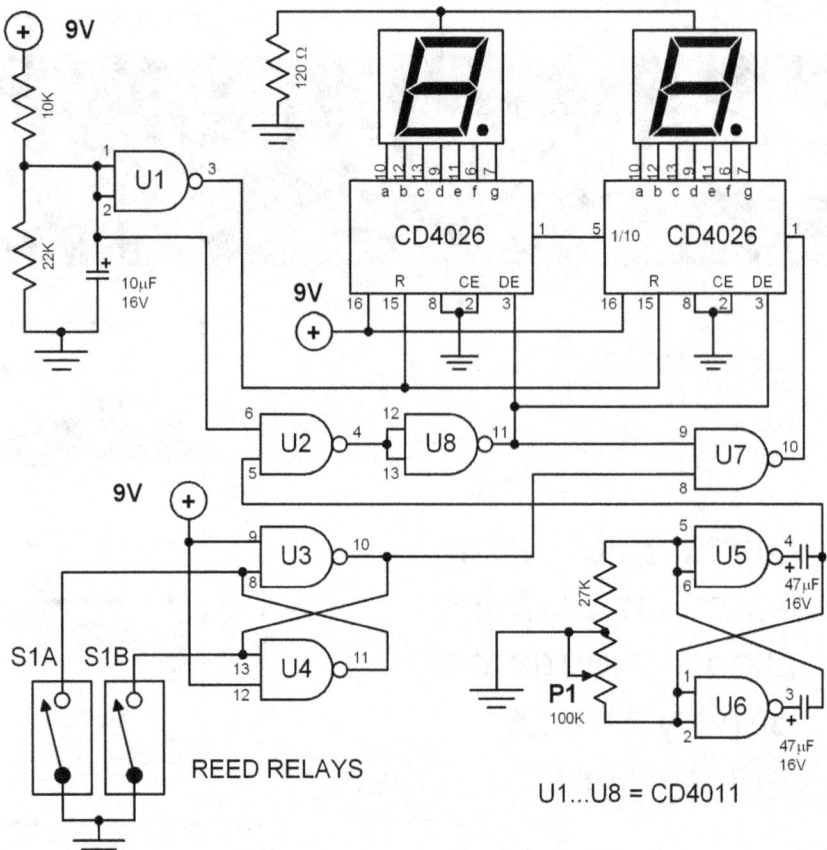

Diagram 98.0 Digital Bike Tacho

This tachometer uses two reed switches to get the speed information of the bicycle. The reed switches are installed near the rim of the wheel where permanent magnets pass by. The permanent magnets are attached to the wheelspokes and activate the reed switches everytime they pass by it. The speed is digitally displayed.

The tachometer circuit works according to this principle; the pulses created by the reed contacts are counted within a certain time interval. The resulting count is then displayed and represents the speed of the bike. Two 4026 ICs are used to count the pulses, decode the counter and control two 7-segment LED display. RS flip-flops U3 and U4 function as anti-bounce. The pulses arrive at the counter's input through gate U7. The measuring period is determined by the monostable multivibrator U5/U6 and can be adjusted through potentiometer P1 so that the tacho can be calibrated. The circuit U1/U2 resets the counters.

Since batteries are used to power the circuit, it is not practical to support the continuous display of speed information. This circuit is not continously active. The circuit is activated only after a button is pressed. At least three permanent magnets must be installed on the wheel. The circuit can be calibrated with the help of another precalibrated tachometer.

99 AUTO CAR WARNING

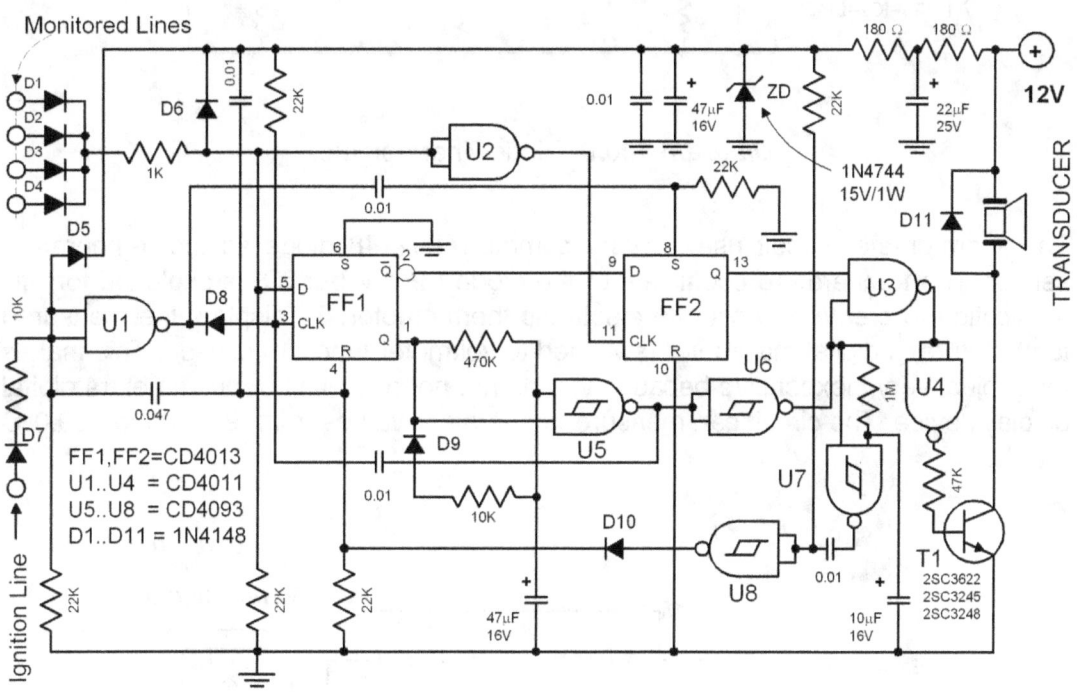

Diagram 99.0 Auto Car Warning

This circuit gives an acoustic signal when the car driver has forgotten to switch off one or more electrical switches which it monitors after the ignition switch is turned off. It has a time delay to give the driver enough time to switch off all these important switches.

100 DIODE THERMOMETER

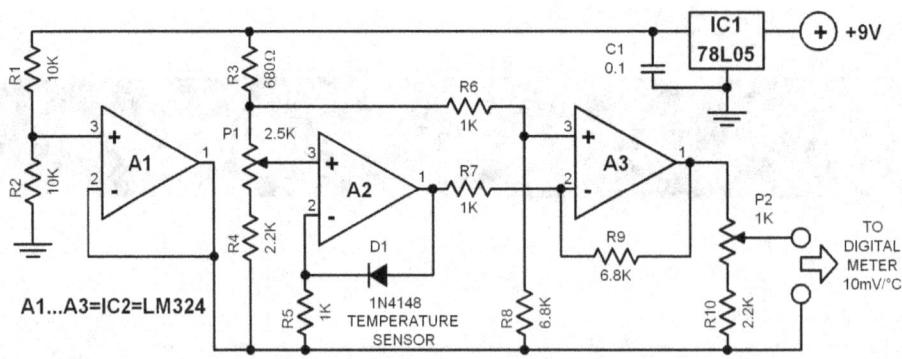

Diagram 100.0 Diode Thermometer

This thermometer circuit uses a very common 1N4148 diode as the temperature sensor. The temperature coefficient of the diode (-2 mV per °C) is exploited for this application to create an accurate electronic thermometer. To display the measured temperature, a digital multimeter (switched to voltmeter function) is used. This makes the project very inexpensive because you do not need to build or buy an extra digital display device. The circuit can measure temperature values from -9.99 up to +99.0°C

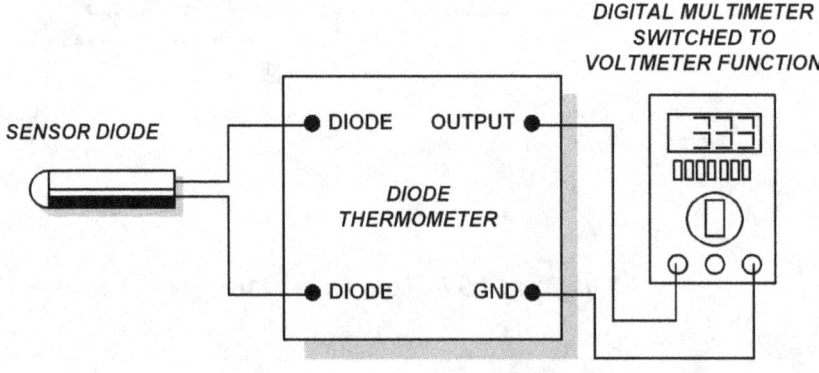

The combination of a diode thermometer module with a digital multimeter that is switched to its voltmeter function makes a functional digital voltmeter.

Calibration:

To set the minimum level (0°C), place the diode in a glass of water filled with crushed ice - check the temperature first with a normal thermometer- wait until the thermometer shows zero degrees centigrade. Set P1 so that the digital voltmeter will display 000 when the diode senses zero degree centigrade.

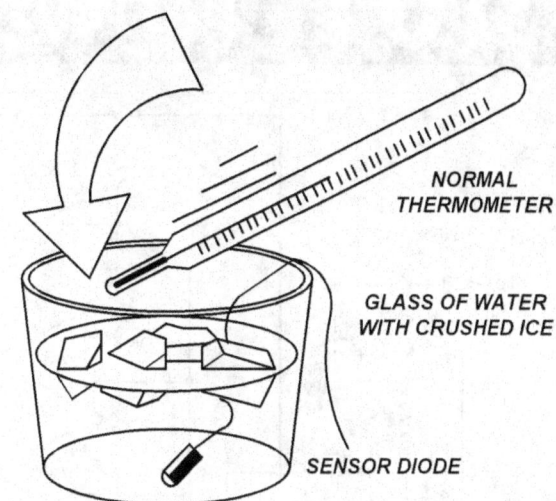

NORMAL THERMOMETER

GLASS OF WATER WITH CRUSHED ICE

SENSOR DIODE

To set the maximum level (100 °C), place the diode sensor into a boiling water (use a thermometer to be sure that the temperature is exactly 99.9°C) and adjust P2 so that the digital meter exactly displays 99.9.

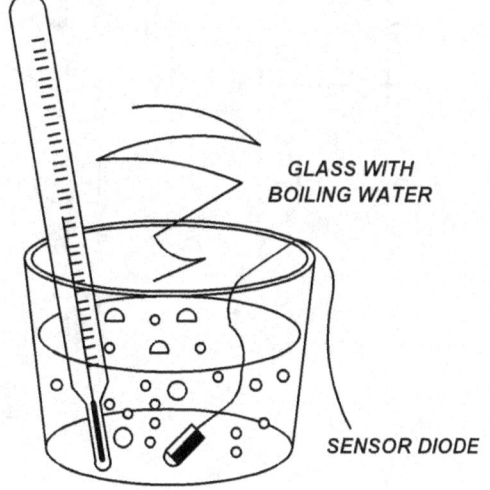

GLASS WITH BOILING WATER

SENSOR DIODE

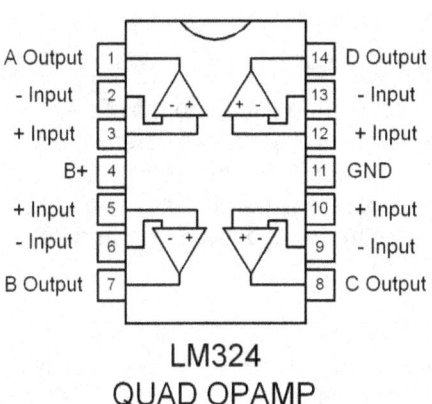

A Output	1		14	D Output
- Input	2		13	- Input
+ Input	3		12	+ Input
B+	4		11	GND
+ Input	5		10	+ Input
- Input	6		9	- Input
B Output	7		8	C Output

LM324
QUAD OPAMP

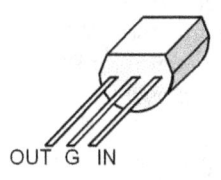

OUT G IN

78L05

101 BATTERY MONITOR

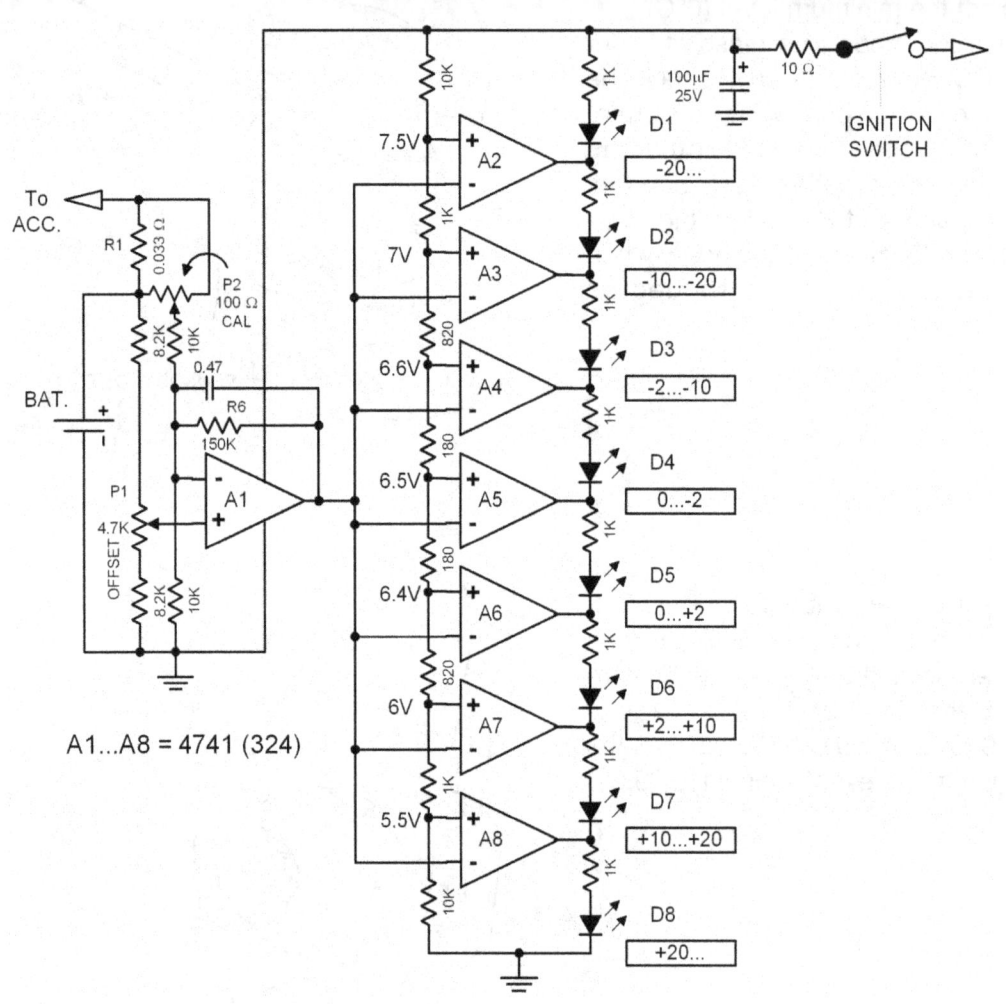

Diagram 101.0 Battery Monitor

The current at resistor R1 produces a voltage drop which is proportional to the current value (Ohm's law forgotten?...). This voltage drop is fed to the inverting input of the differential amplifier IC A1. Potentiometer P2 serves as a calibrator. The IC A1 controls the buffer ICs A2 up to A8. An optical scale composed of LEDs D1 up to D8 displays the electrical current value. By using the calibration potentiometer P2, the output level of IC A1 can be set between 6.5 and 6.6 volts so that LEDs D1...D5 will light.

During discharging, the voltage drop at R1 increases and the LEDs D5...D8 show the discharge. When the current flows through R1 in the reverse direction, the decreasing voltage drop will be displayed by the LEDs. The given component values are calibrated at 6.5 volts for a battery voltage of 13 volts.

102 CAR ANTENNA AMPLIFIER

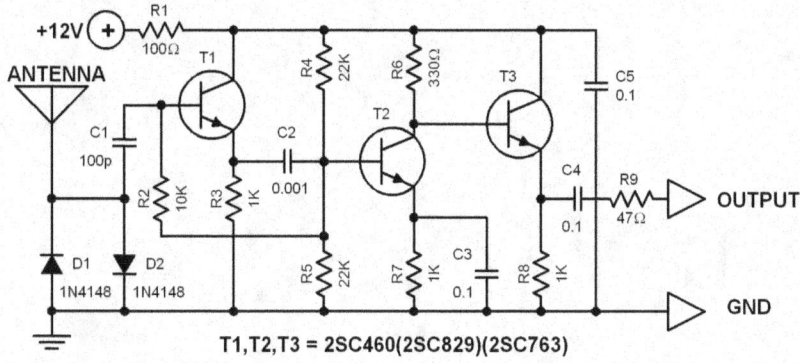

Diagram 102.0 Car Antenna Amplifier

This antenna amplifier is specially designed to boost the weak signals captured by the car antenna. It has a high input impedance to avoid high signal losses. It has also a low noise character.

The amplifier circuit can be used up to 70 MHz. The gain is around 30 dB and the input impedance at 30MHz is around 10K. The amplifier must be mounted directly at the base of the antenna to avoid signal losses caused by the capacitive character of the coaxial cable. Obviously, this antenna amplifier can be used for non-mobile receivers. If you intend to install this circuit in your outdoor mounted antenna, make sure that it is housed in a water proof case. Use this circuit only for receiver antennas. Transmitting through it will damage the components.

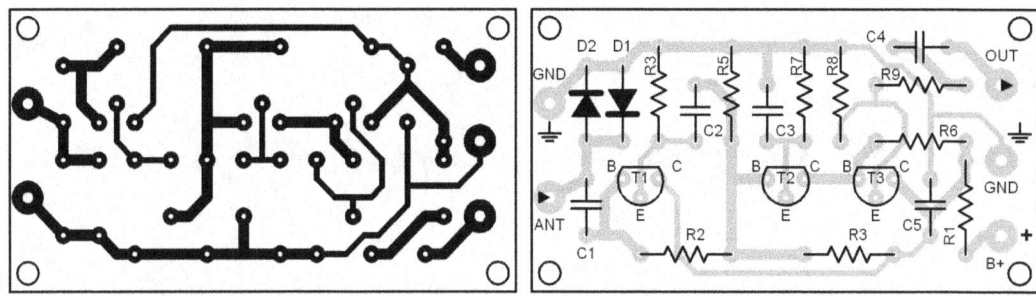

Figure 102.0 Printed Circuit Layout   **Figure 102.1** Parts Placement

This page is intentionally blank.

APPENDICES

Electrical Specifications of the
transistors used in the projects **152**

Diode Specifications **157**

Zener Diode Specifications **158**

Power FETs **160**

Small Signal FETs **161**

Package Information of FETs **162**

Three-terminal Voltage Regulators **163**

Package Information of 3-Terminal
Voltage Regulators **164**

Printed Circuit board Layouts **165**

Specifications of the transistors used in the projects

Descriptive Part of the Table:

Type
The original type designation has been taken over directly from the manufacturers, with the abbreviation of the manufacturer added in brackets only in those cases in which different manufacturers used the same type designation.

Mat.
The materials used are abbreviated as follows:

Ge	Germanium
Mos	MOS technology (metal oxide silicon)
Si	Silicon
V-MOS	Vertical MOS technology

Pol.
The polarities used are abbreviated as follows:

npn	NPN structure
n-ch	N channel type (FET)
n-p	More than one transistor with different polarities in one case
pnp	PNP structure
p-ch	P channel type (FET)

Abbreviations used in the following table:

A	Antenna amplifer		FET	Field-effect transistor
AGC	Regulating steps		FET-depl.	Field-effecttransistor, depletion type
AF	AF range		FET-enh.	Field-effect transistor, enhancement type
AM	AM range		FM	FM range
CATV	Broad band cable amplifier		fs	Fast switch
CB	CB-radio		HD	Horizontal deflection
CTV	Colour television application		hi-rel	high reliability
chop	Chopper		Idss	Drain source short-circuit current (FET)
Darl	Darlington transistor		IF	IF applications
dg	Dual Gate (FET)		in	Input stages
double	Paired types		iso	insulated
dr	Driver stages		ln	Low noise
dual	Dual transistor (differential amplifier)		min	Miniaturised version
			mix	Mixer stages
end	Final stages		nixie	Digital display tube

osc	Oscillator stages	Ugs	Gate source voltage
pow	Power stages	UHF	UHF range > 250MHz
radiation	Aerospace applications (radiation-proof)	uni	Universal type
		Up	Pinch-off voltage (FET)
RF	RF range	VD	Vertical deflection
s	Switch	VHF	VHF range 100-250
SMP	Switch-mode power supply	MHz	
SSB	Single side-band operation	Vid	Video output stages
Stabi	Stabilisation	+Diode, +di	With integrated diode
sym	Symmetrical types	../..ns	turn-on/turn-off time
TV	Television applications		

Data Part of the Table:

In the case of the ratings, either average values are quoted (< = max.) or lower (> = min.) guaranteed values. As a rule apply at 25°C, unless otherwise indicated.

Uc

With transistors, the usual situation is for U_{CBO}(colletor base reverse bias) to be quoted, or U_{CEO} and U_{CEO} (collector emitter reverse bias). With FETs, U_{DS} (drain source voltage) is always quoted.

Ic

With transistors, I_C (collector current) is always quoted. If this is followed by (ss) in brackets, I_{CM} is quoted, i.e. the peak value of the collector current. With FETs, I_D (drain current) is always quoted.

Ptot

As a rule, the total leakage power Ptot is quoted, with RF types we always quote the RF output power P_Q, with corresponding frequency in brackets.

Amplification

The DC current gain B(h_{FE}) or the short-circuit current gain ß(h_{fe}) are always quoted as guaranteed values.

fT

The transition frequency is always qouted in MHz.

Specifications of the transistors used in the projects

Type	Mat.	Pol.	Description	UC [Vmax]	IC [Amax]	Ptot [Wmax]	Current Gain	fT [MHz]
MJ3001	Si	npn	Darl+diode,pow	60	10.00	150.00($25°C)	>10	
MJE243	Si	npn	AF-s-pow	100	4.00	1.50($25°C)	40.120	>40.00
MJE244	Si	npn	AF-s-pow	100	4.00	1.50($25°C)	>25	>40.00
MJE253	Si	npn	AF-s-pow	100	4.00	1.50($25°C)	40-120	>40.00
MJE4350	Si	pnp	AF-end,s-pow	100	16.00	125.00($25°C)	15	>1.00
MJE5170	Si	pnp	uni-pow	120	6.00	2.00($25°C)	15-100	>1.00
MJE5180	Si	npn	uni-pow	120	6.00	2.00($25°C)	15-100	>1.00
MPF102	Si	n-ch	FET,VHF-in,sym,mix 25V,Idss>2mA,Up<V					
MPF106	Si	n-ch	FET,VHF 25V,Idss>4mA,Up<8V					
MPS-A29	Si	npn	Darl	100	0.50	1.50($25°C)	>10	>125.00
2N708	Si	npn	s	40/15	0.20	1.20(25°C)	>15	480.00
2N1711	Si	npn	uni	75	0.50	3.00(25°C)	75	>70.00
2N1889	Si	npn	AF-s	100/60	0.50	3.00(25°C)	40-120	>50.00
2N1890	Si	npn	AF-s	100/60	0.50	3.00(25°C)	100-300	>60.00
2N1990	Si	npn	nixie	100	1.00	2.00(25°C)	>25	
2N2102	Si	npn	AF-s	120/65	1.00	5.00(25°C)	40-120	>120.00
2N2222	Si	npn	ini	0		1.80(25°C)		
2N2368	Si	npn	fs	40/15	0.20	1.20(25°C)	20-60	>400.00
2N2369	Si	npn	fs	40/15	0.20	1.20(25°C)	40-120	>500.00
2N2905	Si	pnp	uni	60/40	0.60	3.00(25°C)	100-300	>200.00
2N2904	Si	pnp	uni	60/40	0.60	3.00(25°C)	40-120	>200.00
2N3019	Si	npn	uni	140/80	1.00	5.00(25°C)	100-300	>100.00
2N3020	Si	npn	uni	140/80	1.00	5.00(25°C)	40-120	>80.00
2N3055	Si	npn	AF-s-pow	100/60	15.00	115.00($25°C)	20-70	>2.50
2N3109	Si	npn	AF-s	80/40	1.00	5.00(25°C)	100-300	>70.00
2N3110	Si	npn	AF-s	80/40	1.00	5.00(25°C)	40-120	>60.00
2N3367	Si	n-ch	FET,uni,In	40V,Idss>0.5mA,Up<2.5V				
2N3370	Si	n-ch	FET,uni,In	40V,Idss>0.1mA,Up3.2V				
2N3454	Si	n-ch	FET,uni	50V,Idss>0.05mA,Up<2.3V				
2N3819	Si	n-ch	FET,VHF,uni,sym	25V,Idss>2mA,Up<8V				
2N3823	Si	n-ch	FET,VHF,In	30V,Idss>4mA,Up<8V				
2N3903	Si	npn	uni	60/40	0.20	1.50(25°C)	50-150	>250.00
2N3904	Si	npn	uni	60/40	0.20	1.50(25°C)	100-300	>300.00
2N3905	Si	pnp	uni	40	0.20	1.50(25°C)	50-150	>200.0
2N3906	Si	pnp	uni	40	0.20	1.50(25°C)	100-300	>250.00
2N4118	Si	n-ch	FET,uni	40V,Idss>0.08mA,Up<3V				
2N5294	Si	npn	AF-s-pow	80/70	4.00	1.80($25°C)	30-120	>0.80
2N5397	Si	n-ch	FET,VHF/UHF	25V,Idss>10mA,Up<6V				
2N5398	Si	n-ch	FET,VHF/UHF	25V,Idss>5mA,Up<6V				
2N5486	Si	n-ch	FET,VHF/UHF	25V,Idss>8mA,Up<6V				
2N6038	Si	npn	Darl+diode,pow	60	4.00	1.50($25°C)	>10	>25.00
2N6039	Si	npn	Darl+diode,pow	80	4.00	1.50($25°C)	>10	>25.00
2N6283	Si	npn	Darl+diode,pow	80	20.00	160.00($25°C)	>10	>4.00
2N6284	Si	npn	Darl+diode,pow	100	20.00	160.0($25°C)	>10	>4.00
2N6412	Si	npn	AF-s-pow	60/40	4.00	15.00($25°C)	>5	>50.00
2N6414	Si	pnp	AF-s-pow	80/60	4.00	15.00($25°C)	>5	>50.00
2SA511	Si	pnp	AF/RF/s	90/80	1.50	8.00(25°)	30-150	60.00
2SA597	Si	pnp	RF-s	50/40	1.00	6.00($25°C)	10-250	400.00
2SA761	Si	pnp	uni	110	2.00	6.30($25°)	50-240	80.00
2SA970	Si	pnp	AF,In	120	0.10	0.30(25°C)	200-700	100.0

Specifications of the transistors used in the projects

Type	Mat.	Pol.	Description	UC [Vmax]	IC [Amax]	Ptot [Wmax]	Current Gain	fT [MHz]
2SA1016	Si	pnp	uni,ln	120/100	0.05	0.40(25°)	160-960	110.00
2SA1123	Si	pnp	uni,ln	150	0.05	0.7(25°)	65-450	200.00
2SA1136	Si	pnp	AF-in,ln	120/100	0.10	0.30(25°C)	120-560	90.00
2SA1137	Si	pnp	AF-in,on	80	0.10	0.30(25°C)	120-560	90.00
2SA1141	Si	pnp	AF/Rf-pow	115	10.00	2.00($25°C)	100	80.00
2SA1285	Si	pnp	uni	120	0.20	0.90(25°C)	150-800	200.00
2SA1285A	Si	pnp	uni	150	0.10	0.90(25°C)	150-500	200.00
2SA1515	Si	pnp	uni	40/32	1.00	0.50(25°C)	82-390	150.00
2SA1705	Si	pnp	AF,s	60/50	1.00	0.90(25°C)	>30	150.00
2SA1706	Si	pnp	AF-s	60/50	2.00	1.00(25°C)	>40	150.00
2SB633	Si	pnp	AF-s-pow	100/85	6.00	40.00($25°C)	40-320	15.00
2SB764	Si	pnp	uni	60/50	1.00	0.90(25°C)	60-320	150.00
2SB822	Si	pnp	Af-dr/end	40/32	2.00	0.75(25°C)	82-390	100.00
2SB826	Si	pnp	s-pow	60/50	7.00	60.00($25°C)	>30	10.00
2SB867	Si	pnp	AF/s-pow,lo-sat	130/80	3.00	30.00($25°C)	60-260	30.00
2SB868	Si	pnp	AF/s-pow,lo-sat	130/80	4.00	35.00($25°C)	60-260	30.00
2SB869	Si	pnp	AF/s-pow,lo-sat	130/80	5.00	40.00($25°C)	60-260	30.00
2SB870	Si	pnp	AF/s-pow,lo-sat	120/80	7.00	40.00($25°C)	60-260	30.00
2SB874	Si	pnp	AF/s-pow, TV-VD	100/60	2.00	20.00($25°C)	>40	250.00
2SB909	Si	pnp	AF-dr/end	40/32	1.00	1.00(25°C)	82-390	150.00
2SB911	Si	pnp	AF-dr/end	40/32	2.00	1.00(25°C)	82-390	100.0
2SB920	Si	pnp		120/80				
2SB921	Si	pnp		120/80				
2SB1064	Si	pnp	AF-s-pow	60/50	3.00	1.50($25°)	60-320	70.00
2SB1114	Si	pnp	min,uni	20	2.00	2.00($25°C)	135-600	180.00
2SB1116	Si	pnp	uni	60/50	1.00	0.75(25°C)	135-600	120.00
2SB1142	Si	pnp	s-pow	60/50	2.50	10.00(25°C)	>35	140.00
2SB1143	Si	pnp	s-pow	60/50	4.00	10.00(25°C)	>40	150.00
2SB1144	Si	pnp	AF/s-pow,lo-sat	120/100	1.50	10.00(25°C)	>30	100.00
2SB1230	Si	pnp	AF/s-pow,lo-sat	110/100	15.00	100.00($25°C)	50-140	
2SB1231	Si	pnp	AF/s-pow,lo-sat	110/100	25.00	120.00($25°C)	50-140	
2SB1232	Si	pnp	AF/s-pow,lo-sat	110/100	40.00	150.00($25°C)	50-140	
2SC270	Si	npn	s-pow	270/75	5.00	50.00($25°C)	24-40	22.00
2SC460	Si	npn	AM-in/mix/osc	30	0.10	0.20(25°C)	35-200	230.00
2SC696	Si	npn	uni	100/60	3.00	0.75(25°C)	30-173	100.00
2SC763	Si	npn	VHF	25/12	0.02	0.10(25°C)	20-300	>400.00
2SC829	Si	npn	AM/FM-in/mix/osc	30/20	0.03	0.40(25°C)	40-500	230.00
2SC959	Si	npn	uni	120/80	0.70	0.70(25°C)	40-200	100.00
2SC1324	Si	npn	UHF-CATV	35/25	0.15	3.00(25°C)	10-35	
2SC1876	Si	npn	Darl	100/70	0.50	0.80(25°C)	>20	
2SC2124	Si	npn	TV-HD	220/800	2.00	5.00($90°C)	20	4.00
2SC2125	Si	npn	TV-HD	220/800	5.00	50.00($25°C)	8-25	5.00
2SC2270	Si	npn	lo-sat	50/20	5.00	1.00($25°C)	>70	100.00
2SC2334	Si	npn	s-pow,dc-dc conv.	150/100	7.00	40.00($25°C)	>20	
2SC2459	Si	npn	uni	120	0.10	0.20(25°C)	200-700	100.00
2SC2675	Si	npn	AF,ln	80	0.10	0.30(25°C)	180-820	120.00
2SC2724	Si	npn	FM-IF	30/25	0.03	0.20(25°C)	25-300	200.00
2SC3112	Si	npn	AF,ln	50	0.15	0.40(25°C)	600-3600	250.00
2SC3179	Si	npn	AF-pow	80/60	4.00	30.00($25°C)	100	15.00
2SC3245	Si	npn	uni	120	0.10	0.90(25°C)	150-800	200.00

Specifications of the transistors used in the projects

Type	Mat.	Pol.	Description	UC [Vmax]	IC [Amax]	Ptot [Wmax]	Current Gain	fT [MHz]
2SC3245A	Si	npn	uni	150	0.10	0.90(25°C)	400-800	200.00
2SC3248	Si	npn	uni	180	0.10	0.90(25°C)	150	130.00
2SC3358	Si	npn	UHF	20/12	0.10	0.25(25°C)	50-300	7000.00
2SC3420	Si	npn	lo-sat	50/20	5.00	10.00(25°C)	>70	100.00
2SC3622	Si	npn	AF-s,hi-beta	60/50	0.15	0.25(25°C)	1000-3200	250.00
2SC4308	Si	npn	VHF-A	30/20	0.30	0.60(25°C)	50-200	2500.00
2SD386	Si	npn	TV-VD	200/120	3.00	1.75($25°C)	40-320	8.00
2SD406	Si	npn	Darl	100	2.00	15.00(25°C)	>2000	
2SD613	Si	npn	AF-s-pow	100/85	6.00	40.00($25°C)	40-320	15.00
2SD614	Si	npn	Darl	100/80	3.00	0.80(25°C)	3000	15.00
2SD621	Si	npn	TV_HD	2500/900	3.00	50.00($25°C)	3-15	
2SD628	Si	npn	Darl+diode,pow	100	10.00	80.00($25°C)	>1000	
2SD629	Si	npn	Darl+diode,pow	100	10.00	100.00($25°C)	>1000	
2SD688	Si	npn	Darl,pow	100	1.50	0.80($25°C)	>10	
2SD712	Si	npn	AF-s-pow	100	4.00	30.00($25°C)	55-300	8.00
2SD726	Si	npn	AF-s-pow	100/80	4.00	40.00($25°C)	35-320	10.00
2SD729	Si	npn	Darl+diode,pow	100	20.00	125.00($25°C)	>1000	
2SD781	Si	npn	s-pow,TV-HD	150/60	2.00	1.00(25°C)	150	
2SD826	Si	npn		60/20	5.00	1.00($25°C)	120-560	120.00
2SD838	Si	npn	TV-HD,s-pow	2500/900	3.00	50.00($25°C)	3-15	
2SD892A	Si	npn	Darl	60/50	0.50	0.40(25°C)	>8000	150.00
2SD1049	Si	npn	AF-s-pow	120/80	25.00	80.00($25°C)	>20	
2SD1062	Si	npn	s-pow	60/50	12.00	40.00($25°C)	>30	10.00
2SD1153	Si	npn	Darl	80750	1.50	0.90(25°C)	>40	120.00
2SD1177	Si	npn	AF-pow,TV-HD	100/60	2.00	20.00($25°C)	>40	230.00
2SD1237	Si	npn	s-pow	90/80	7.00	1.75($25°C)	>30	20.00
2SD1238	Si	npn	s-pow	90/80	12.00	80.00($25°C)	>30	20.00
2SD1639	Si	npn	AF-s-pow	100/80	2.20	10.00($25°C)	40-200	
2SD1684	Si	npn	AF/s-pow,lo-sat	120/100	1.50	10.00(25°C)	>30	120.00
2SD1685	Si	npn	AF/s-pow,lo-sat	60/20	5.00	10.00(25°C)	>95	120.00
2SD1691	Si	npn	AF-s-pow	60	5.0	20.00(25°C)	100-400	
2SD1840	Si	npn	AF/s-pow,lo-sat	110/100	15.00	100.00($25°C)	50-140	
2SD1841	Si	npn	AF/s-pow,lo-sat	110/100	25.00	120.00($25°C)	50-140	
2SD1842	Si	npn	AF/s-pow,lo-sat	110/100	40.00	150.00($25°C)	50-140	
2SD2116	Si	npn	Darl	80/50	0.70	1.00(25°C)	>40	
2SD2117	Si	npn	Darl	80/50	1.50	1.00(25°C)	>30	
2SD2213	Si	npn	Darl,AF	150/80	1.50	0.90(25°C)	>10	
2SJ165	V-MOS	p-ch	FET-enh.,	50V,0.1A,0.25W				
2SK422	V-MOS	n-ch	FET-enh.	60v,0.7A,0.9W,17/12ns				
2SK423	V_MOS	n-ch	FET-enh.	100V,0.5A,0.9W,15/20ns				
3N140	MOS	n-ch	FET-depl.,dg,FM/VHF-in	20V,Idss>5mA				
3N225	MOS	n-ch	FET-depl.,dg, UHF	25V,Idss>1mA,Up<4V				
3SK35	MOS	n-ch	FET-depl.,dg,VHF	20V,Idss>3mA,Up<4V				
3SK37	MOS	n-ch	FET-depl.,dg,VHF	20V,Idss>4mA,Up<3V				
3SK45	MOS	n-ch	FET-depl.,dg,VHF	22V,Idss>4mA,Up<3V				
3SK61	MOS	n-ch	FET-depl.,dg,VHF	20V,Idss>4mA,Up<3V				
3SK72	MOS	n-ch	FET-depl.,dg,VHF	20V,Idss>2.5mA,Up<3V				
3SK77	MOS	n-ch	FET-depl.,dg,VHF	20V,Idss>3mA,Up<2.5V				
3SK85	MOS	n-ch	FET-depl.,dg,VHF	20V,Idss>4mA,Up<3V				

SEMICONDUCTOR DIODE SPECIFICATIONS

* RFR = Rectifier, Fast Recovery

Device	Type	Material	Peak Inverse Voltage, PIV (Volts)	Average Rectified Current Forward (Reverse) IO (A) (IR(A))	Peak Surge Current, IFSM 1 sec. @ 25ºC (A)	Average Forward Voltage, VF (Volts)
1N34	Signal	Germanium	60	8.5 m (15.0μ)		1.0
1N34A	Signal	Germanium	60	5.0 m (30.0μ)		1.0
1N67A	Signal	Germanium	100	4.0 m (5.0μ)		1.0
1N191	Signal	Germanium	90	5.0 m	1.0	
1N270	Signal	Germanium	80	0.2 (100μ)		1.0
1N914	Fast Switch	Silicon (Si)	75	75.0 m (25.0 n)	0.5	1.0
1N1184	RFR	Si	100	35 (10 m)		1.7
1N2071	RFR	Si	600	0.75 (10.0μ)		0.6
1N3666	Signal	Germanium	80	0.2 (25.0μ)		1.0
1N4001	RFR	Si	50	1.0 (0.03 m)		1.1
1N4002	RFR	Si	100	1.0 (0.03 m)		
1N4003	RFR	Si	200	1.0 (0.03 m)		1.1
1N4004	RFR	Si	400	1.0 [0.03 m)		1.1
1N4005	RFR	Si	600	1.0 (0.03 m)		1.1
1N4006	RFR	Si	800	1.0 (0.03 m)		1.1
1N4007	RFR	Si	1000	1.0 (0.03 m)		1.1
1N4148	Signal	Si	75	10.0 m (25.0 n)		1.0
1N4149	Signal	Si	75	10.0 m (25.0 n)		1.0
1N4152	Fast Switch	Si	40	20.0 m (0.05μ)		0.8
1N4445	Signal	Si	100	0.1 (50.0 n)		1.0
1N5400	RFR	Si	50	3.0	200	
1N5401	RFR	Si	100	3.0	200	
1N5402	RFR	Si	200	3.0	200	
1N5403	RFR	Si	300	3.0	200	
1N5404	RFR	Si	400	3.0	200	
1N5405	RFR	Si	500	3.0	200	
1N5406	RFR	Si	600	3.0	200	
1N5767	Signal	Si		0.1 (1.0μ)		1.0
ECG5863	RFR	Si	600	6	150	0.9

ZENER DIODES SPECIFICATIONS

Zener Voltage (Volts)	Power (Watts)							
	0.25	0.4	0.5	1.0	1.5	5.0	10.0	50.0
1.8	1N4614							
2.0	1N4615							
2.2	1N4616							
2.4	1N4617	1N4370,A	1N4370,A,1N5221,B 1N5985,B					
2.5			1N5222B					
2.6	1N702,A							
2.7	1N4618	1N4371,A	1N4371,A,1N5223,B 1N5839, 1N5986					
2.8			1N5224B					
3.0	1N4619	1N4372,A	1N4372,1N5225,B 1N5987					
3.3	1N4620	1N746,A 1N764 A 1N5518	1N746A 1N5226,B 1N5988	1N3821 1N4728,A	1N5913	1N5333,B		
3.6	1N4621	1N747,A 1N5519	1N747A 1N5227,B,1N5989	1N3822 1N4729,A	1N5914	1N5334,B		
3.9	1N4622	1N748,A 1N5520	1N748A,1N5228,B 1N5844, 1N5990	1N3823 1N4730,A	1N5915	1N5335,B	1N3993A	1N4549,B 1N4557,B
4.1	1N704,A							
4.3	1N4623	1N749,A 1N5521	1N749,A 1N5229,B 1N5845,1N5991	1N3824 1N4731 ,A	1N5916	1N5336,B	1N3994,A	1N4550,B 1N4558,B
4.7	1N4624	1N750,A 1N5522	1N750A ,1N5230,B 1N5846, 1N5992	1N3825 1N4732,A	1N5917	1N5337,B	1N3995,A	1N4551,B 1N4559,B
5.1	1N4625 1N4689	1N751 A 1N5523	1N751A, 1N5231,B 1N5847,1N5993	1N3826 1N4733	1N5918	1N5338,B 1N4560,B	1N3996,A	1N4552,B
5.6	1N708A 1N4626	1N752,A 1N5524	1N752,A,1N5232,B 1N5848, 1N5994	1N3827 1N4734,A	1N5919	1N5339,B 1N4561,B	1N3997,A	1N4553,B
5.8	1N706A	1N762						
6.0				1N5233B 1N5849			1N5340,B	
6.2	1N709,1N4627 MZ605,MZ610 MZ620,MZ640	1N753,A 1N821,3,5, 7,9; A	1N753,A 1N5234,B, 1N5850 1N5995	1N3828,A 1N4735,A	1N5920	1N5341,B 1N4562,B	1N3998,A	1N4554,B
6.4	1N4565-84,A							
6.8	1N4099	1N754,A 1N957,B 1N5526	1N754,A 1N757,B 1N5235,B 1N5851 1N5996	1N3016,B 1N3829 1N4736,A	1N3785 1N5921	1N5342,B	1N2970,B 1N3999,A	1N2804B 1N3305B 1N4555, 1N4563
7.5	1N4100	1N755,A 1N958,B 1N5527	1N755A,1N958,B 1N5236,B, 1N5862 1N5997	1N3017,A,B 1N3830 1N4737,A	1N3786 1N5922	1N5343,B 1N4000,A 1N4556,	1N2971,B 1N3306,B	1N2805,B 1N4564
8.0	1N707A							
8.2	1N712A 1N4101	1N756,A 1N959,B 1N5528	1N756,A 1N959,B,1N5237,B 1N5853 ,1N5998	1N3018,B 1N4738,A	1N3787 1N5923	1N5344,B	1N2972,B	1N2806,B 1N3307,B
8.4		1N3154-57,A 1N3155-57	1N3154,A					
8.5	1N4775-84,A		1N5238,B,1N5854					
8.7	1N4102					1N5345,B		
8.8		1N 764						
9.0		1N764A	1N935-9;A,B					

ZENER DIODES SPECIFICATIONS

Zener Voltage (Volts)	Power (Watts)							
	0.25	0.4	0.5	1.0	1.5	5.0	10.0	50.0
9.1	1N4103	1N757,A 1N960,B 1N5529	1N757,A, 1N960,B 1N5239,B, 1N5855 1N5999	1N3019,B 1N4739,A	1N3788 1N5924	1N5346,B	1N2973,B	1N2807,B 1N3308,B
10.0	1N4104	1N758,A 1N961,B 1N5530,B	1N758,A, 1N961,B 1N5240,B, 1N5856 1N6000	1N3020,B 1N4740	1N3789 1N5925	1N5347,B	1N2974,B	1N2808,B 1N3309,A,B
11.0	1N715,A 1N4105	1N962,B 1N5531	1N962,B,1N5241,B 1N5857, 1N6001	1N3021,B 1N4741,A	1N3790 1N5926	1N5348,B	1N2975,B	1N2809,B 1N3310,B
11.7	1N716,A 1N4106		1N941,A,B					
12.0		1N759,A 1N963,B 1N5532	1N759,A ,1N963,B 1N5242,B, 1N5858 1N6002	1N3022,B 1N4742,A	1N3791 1N5927	1N5349,B	1N2976,B	1N2810,B 1N3311,B
13.0	1N4107	1N964,B 1N5533	1N964,B,1N5243,B 1N5859,1N6003	1N3023,B 1N4743,A	1N3792 1N5928	1N5350,B	1N2977,B	1N2811,B 1N3312,B
14.0	1N4108	1N5534	1N5244B, 1N5860			1N5351,B	1N2978,B	1N2812,B 1N3313,B
15.0	1N4109	1N965,B 1N5535	1N965,B,1N5245,B 1N5861,1N6004	1N3024,B 1N4744A	1N3793 1N5929	1N5352,B	1N2979,A,B	1N2813,A,B 1N3314,B
16.0	1N4110	1N966,B 1N553,B	1N966,B,1N5246,B 1N5862, 1N6005	1N3025,B 1N4745,A	1N3794 1N5930	1N5353,B	1N2980,B	1N2814,B 1N3315,B
17.0	1N4111	1N5537	1N5247,B 1N5863			1N5354,B	1N2981B	1N2815,B 1N3316,B
18.0	1N4112	1N967,B 1N5538	1N967,B 1N5248,B 1N5864, 1N6006	1N3026,B 1N4746,A	1N3795 1N5931	1N5355,B	1N2982,B	1N2816,B 1N3917,B
19.0	1N4113	1N5539	1N5249,B 1N5865			1N5356,B	1N2983,B	1N2817,B 1N3318,B
20.0	1N4114	1N968,B 1N5540	1N968,B,1N5250,B 1N5866, 1N6007	1N3027,B 1N4747,A	1N3796 1N5932,A,B	1N5357,B	1N2984,B	1N2818,B 1N3319,B
22.0	1N4115	1N959,B 1N5541	1N969,B,1N5241,B 1N5867, 1N6008	1N3028,B 1N4748,A	1N3797 1N5933	1N5358,B	1N2985,B	1N2819,B 1N3320,A,B
24.0	1N4116	1N5542 1N9701B	1N970,B,1N5252,B 1N586,1N6009	1N3029,B 1N4749,A	1N3798 1N5934	1N5359,B	1N2986,B	1N2820,B 1N3321,B
25.0	1N4117	1N5543	1N5253,B 1N5869			1N5360,B	1N2987B	1N2821,B 1N3322,B
27.0	1N4118	1N971,B	1N971,1N5254,B 1N5870,1N6010	1N3030,B 1N4750,A	1N3799 1N5935	1N5361,B	1N2988,B	1N2822B 1N3323,B
28.0	1N4119	1N5544	1N5255,B,1N5871			1N5362,B		
30.0	1N4120	1N972,B 1N5546	1N972,B,1N5256,B 1N5872,1N6011	1N3031,B 1N4751,A	1N3800 1N5936	1N5363,B	1N2989,B	1N2823,B 1N3324,B
33.0	1N4121	1N973,B 1N5546	1N973,B,1N5257,B 1N5873,1N6012	1N3032,B 1N4752,A	1N3801 1N5937	1N5364,B	1N2990,A,B	1N2824,B 1N3325,B
36.0	1N4122	1N974,B	1N974,B,1N5258,B 1N5874,1N6013	1N3033,B 1N4753,A	1N3802 1N5938	1N5365,B	1N2991,B	1N2825,B 1N3326,B
39.0	1N4123	1N975,B	1N975,B, 1N5259,B 1N5875 ,1N6014	1N3034,B 1N4754,A	1N3803 1N5939	1N5366,B	1N2992,B	1N2826,B 1N3327,B
43.0	1N4124	1N976,B	1N976,B,1N5260,B 1N5876,1N6015	1N3035,B 1N4755,A	1N3804 1N5940	1N5367,B	1N2993,A,B	1N2827,B 1N3328,B
45.0			1N2994B	1N2828B 1N3329B				

POWER FETs

Device No.	Type	Max. Diss. (W)	Max. VDS (Volts)	Max ID (A)*	Gfs mmhos (typ.)	Input Ciss (pF)	Output Coss (pF)	Approx. Upper Freq. (MHz)	Case	Pack-Type Mnfr.	General applications age/ Mnfr.
DV1202S	N-Chan.	10	50	0.5	100k	14	20	500	.380 SOE	1/S	RF power amp., oscillator
DV1202W	N-Chan.	10	50	0.5	100k	14	20	500	C-220	5/S	RF power amp., oscillator
DV1205S	N-Chan.	20	50	1	200k	26	38	500	.380 SOE	1/S	RF power amp., oscillator
DV1205W	N-Chan.	20	50	1	200k	26	98	500	C-220	5/S	RF power amp., oscillator
2SK133	N-Chan.	100	120	7	1M	600	350	1	TO-3	6/H	AF pwr. amp., switch (complem to 25J48)
2SK134	N-Chan.	100	140	7	1M	600	350	1	TO-3	6/H	AF pwr. amp., switch (complem to 25J49)
2SK135	N-Chan.	100	160	7	1M	600	350	1	TO-3	6/H	AF pwr. amp., switch (complem to 25J50)
2SJ48	P-Chan.	100	120	7	1M	900	400	1	TO-3	6/H	AF pwr. amp., switch (complem to 2SK133)
2SJ49	P-Chan.	100	140	7	1M	900	400	1	TO-3	6/H	AF pwr. amp., switch (complem to 2SK134)
2SJ50	P-chan.	100	160	7	1M	900	400	1	TO-3	6/H	AF pwr. amp., switch (complem to 2SK135)
VMP4	N-Chan.	25	60	2	170K	32	4.8	200	.380 SOE	1/S	VHF pwr. amp., rcvr front end (rf amp., mixer).
VN10KM	N-Chan.	1	60	0.5	100K	48	16	-	TO-92	2/S	High-speed line driver, relay driver, LED stroke driver
VN64GA	N-Chan.	80	60	12.5	150K	700	325	30	TO-3	3/S	Linear amp., power-supply switch, motor control
VN66AF	N-Chan.	15	60	2	150K	50	50	-	TO-202	4/S	High-speed switch, HF linear amp., audio amp. line driver.
VN66AK	N-Chan.	8.3	60	2	250K	93	6	100	TO-39	7/S	RF pwr.amp.,high-current analog switching
VN67AJ	N-Chan.	25	60	2	250K	33	7	100	TO-3	3/S	RF pwr.amp.,high-current switching
VN89AA	N-Chan.	25	80	2	250K	50	10	100	TO-3	3/S	High-speed switching,HF linear amps., line drivers.
IRF100	N-Chan.	125	80	16	300K	900	25	-	TO-3	3/S	High-speed switching,audio inverters.
IRF101	N-Chan.	125	60	16	300K	900	25	-	TO-3	3/S	Same as IRF100
Legend:	* 25ºC (case)				S = M/A-COM H = Hitachi		IR = International Rectifier. Mnfr = Manufacturer				

SMALL-SIGNAL FETs

Device No.	Type	Max. Diss. (mW)	Max. V$_{DS}$ (Volts)	Max. I$_D$ (mA)*	Min G$_{fs}$ (mS)	Input C (pF)	V$_{GS(off)}$ (volts)	Upper Freq.(MHz)	Noise Figure	Case Type (typ)	/Mnfr.	General applications
2N4416	N-JFET	300	30	-15	4.5K	4	-6	450	400 MHz 4 dB	TO-72	1/S,M	VHF/UHF/RF amp.mix., osc.
2N5484	N-JFET	310	25	30	2.5K	5	-3	200	200 MHz 4 dB	TO-92	2/M	VHF/UHFamp,mix., osc.
2N5485	N-JFET	310	25	30	3.5K	5	-4	400	400 MHz 4 dB	TO-92	2/S	VHF/UHF/RF amp.mix., osc.
3N200	N-Dual-Gate MOSFET	330	20	50	10K	4-8.5	-6	500	400 MHz 4.5 dB	TO-72	3/R	VHF/UHF/RF amp.mix., osc.
3N202	N-Dual-Gate MOSFET	360	25	50	8K	6	-5	200	200 MHz 4.5 dB	TO-72	3/S	VHF amp., mixer
MPF102	N-JFET	310	25	20	2K	4.5	-8	200	200 MHz	TO-92	2/N,M	HF/VHF amp.,mix., osc.,
MPF106/ 2N5484	N-JFET	310	25	30	2.5K	5	-6	400	200 MHz 4 dB	TO-92	2/N,M	HF/VHF/UHF amp.,mix.,osc.
40673	N-Dual-Gate MOSFET	330	20	50	12K	6	-4	400	200 MHz 6 dB	TO-72	3/R	HF/VHF/UHF amp. mix., osc.
U300	P-JFET	300	-40	20	8K	-50	+10	-	400 MHz	TO-18	4/S	General Purpose amp.
U304	P-JFET	350	-30	-50		27	+10	-	-	TO-18	4/S	analog switch, chopper
U310	N-JFET	500	30	60	10K	2.5	-6	450	450 MHz 3.2 dB	TO-52	5/S	common-gate VHF/UHF amp.,osc., mixer
		30 300										
U350	N-JFET Quad	1W	25	60	9K	5	-6	100	100 MHz 7 dB	TO-99	6/S	matched JFET doubly bal. mixer
U431	N-JFET Dual	300	25	30	10K	5	-6	100	100 MHz -	TO-99	7/S	matched JFET cascade amp., balanced mixer

* 25°C   S = Siliconix Inc.   R = RCA   N = National Semiconductor   M = Motorola

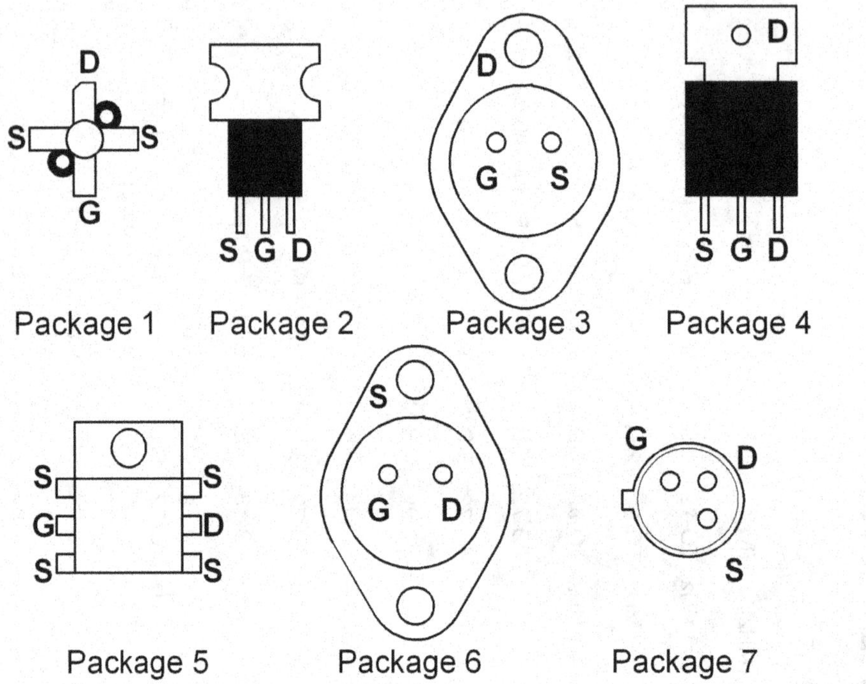

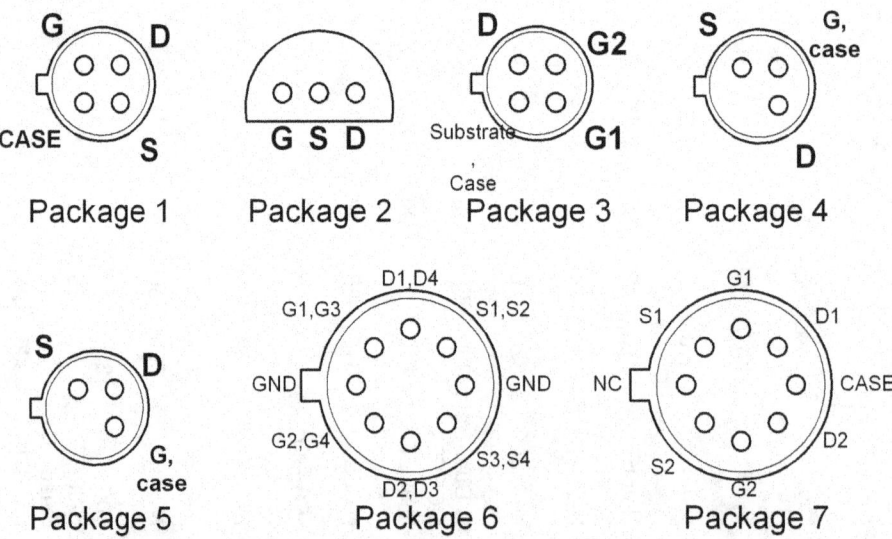

Package Information for Power FETs

Package Information for Small Signal FETs

Three-Terminal Voltage Regulators

* Listed numerically by device

Device	Description	Voltage	Current (Amps)	Package
317	Adj. Pos	+1.2 to +37	0.5	TO-205
317	Adj. Pos	+1.2 to +37	1.5	TO-204,TO-220
317L	Low Current Adj. Pos	+1.2 to +37	0.1	TO-205,TO-92
317M	Med Current Adj. Pos	+1.2 to +37	0.5	TO-220
350	High Current Adj. Pos	+1.2 to +33	3.0	TO-204,TO-220
337	Adj. Neg	-1.2 to -37	0.5	TO-205
337	Adj. Neg	-1.2 to -37	1.5	TO-204,TO-220
337M	Med Current Adj. Neg	-1.2 to -37	0.5	TO-220
309		+5	0.2	TO-205
309		+5	1.0	TO-204
323		+5	9.0	TO-204,TO-220
140-XX	Fixed Pos	**Note #**	1.0	TO-204,TO-220
340-XX			1.0	TO-204,TO-220
78XX			1.0	TO-204,TO-220
78LXX			0.1	TO-205,TO-92
78MXX			0.5	TO-220
78TXX			3.0	TO-204
79XX	Fixed Neg	**Note #**	1 .0	TO-204,TO-220
79LXX			0.1	TO-205,TO-92
79MXX			0.5	TO-220

Legend:

Adj.	= Adjustable
Med	= Medium
Neg	= Negative
Pos	= Positive

Note # - XX indicates the regulated voltage; which may be anywhere from 1.2 volts to 35 volts. For example a 7808 is a positive 8-volt regulator, and a 7912 is a negative 12-volt regulator.

The regulator package may be denoted by an additional suffix, according to the following:

Package	**Suffix**
TO-204 (TO-3)	K
TO-220	T
TO-205 (TO-39)	H,G
TO-92	P,Z

Example:
A 7815K is a positive 15-volt regulator in a TO-204 package. An LM340T-8 is a positive 8-volt regulator in a TO-220 package. In addition, different manufacturers use different prefixes. An LM7812 is equivalent to a µA 7812 or MC7812.

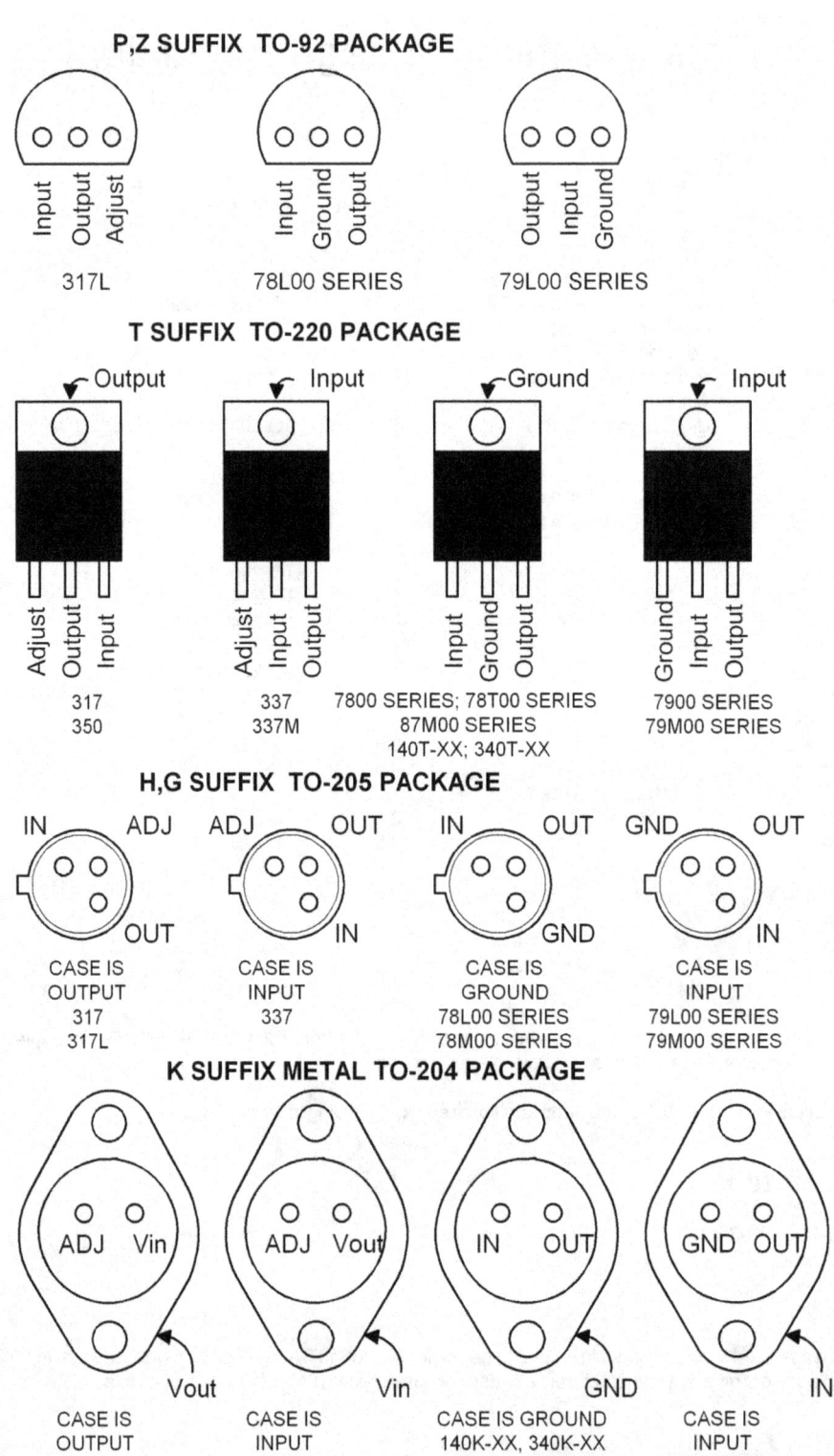

P,Z SUFFIX TO-92 PACKAGE

317L — Input Output Adjust

78L00 SERIES — Input Ground Output

79L00 SERIES — Output Input Ground

T SUFFIX TO-220 PACKAGE

Output — Adjust Output Input
317
350

Input — Adjust Input Output
337
337M

Ground — Input Ground Output
7800 SERIES; 78T00 SERIES
87M00 SERIES
140T-XX; 340T-XX

Input — Ground Input Output
7900 SERIES
79M00 SERIES

H,G SUFFIX TO-205 PACKAGE

IN ADJ OUT
CASE IS
OUTPUT
317
317L

ADJ OUT IN
CASE IS
INPUT
337

IN OUT GND
CASE IS
GROUND
78L00 SERIES
78M00 SERIES

GND OUT IN
CASE IS
INPUT
79L00 SERIES
79M00 SERIES

K SUFFIX METAL TO-204 PACKAGE

ADJ Vin Vout
CASE IS
OUTPUT
317, 350

ADJ Vout Vin
CASE IS
INPUT
337

IN OUT GND
CASE IS GROUND
140K-XX, 340K-XX
309, 7800 SERIES
78T00 SERIES

GND OUT IN
CASE IS
INPUT
79L00 SERIES

PRINTED CIRCUIT BOARD LAYOUTS

All printed circuit board layouts in this collection are once again printed in the following pages. You can either cut out or photocopy these pages to make a separate file for quick reference.

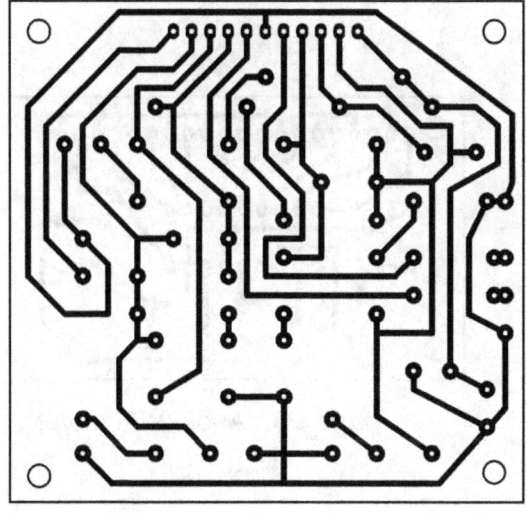

page 12 Stereo Booster

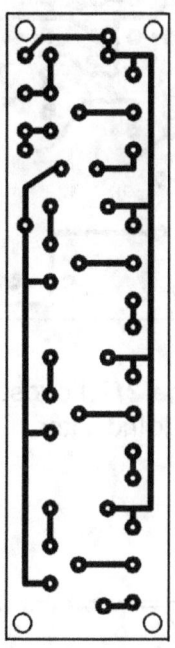

page 15 FET Mixer

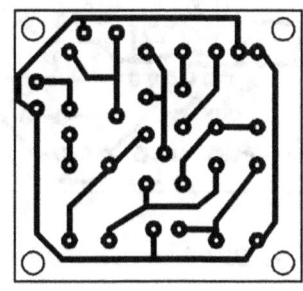

page 18 Low Noise Preamp

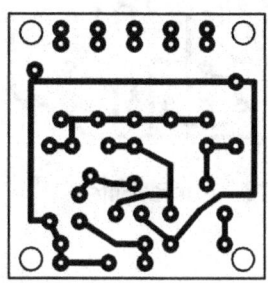

page 19 Audio Mixer

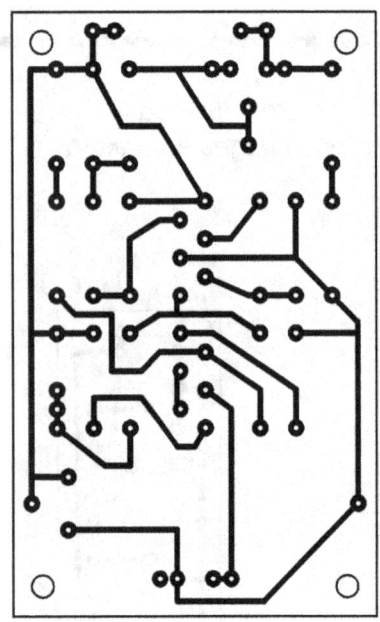

page 21 Loudspeaker
Peak Indicator

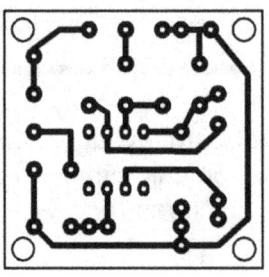

page 24 E-Guitar Preamp

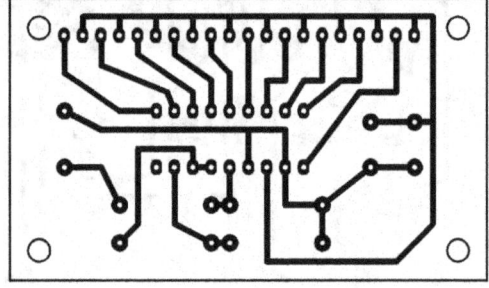

page 28 Audio Wattmeter

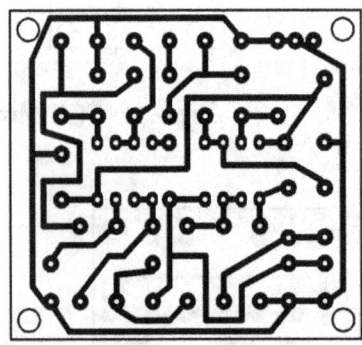

page 23 Audio Filter

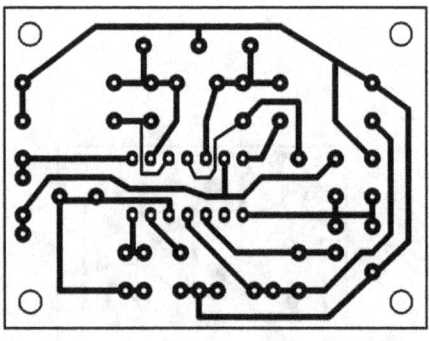

page 32 Microphone Preamp

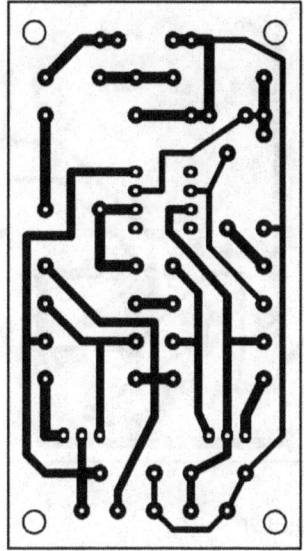

page 33 Headphone Amplifier

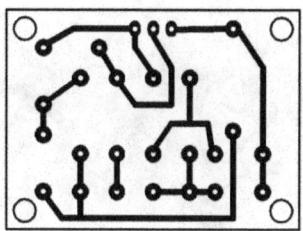

page 37 Motor Speed Regulator

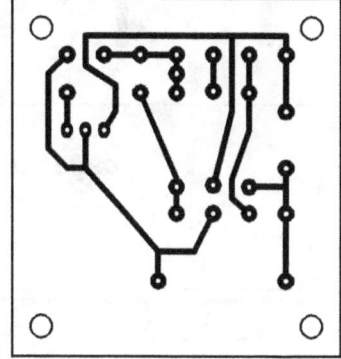

page 37 Motor Speed Regulator #2

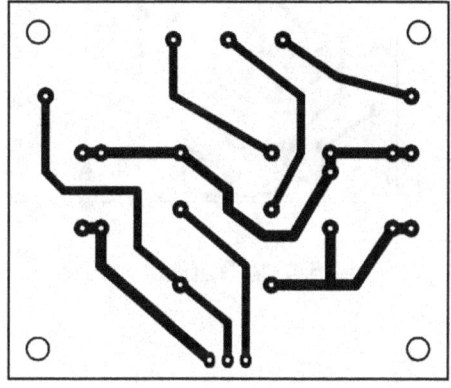

page 42 Motor Regulator

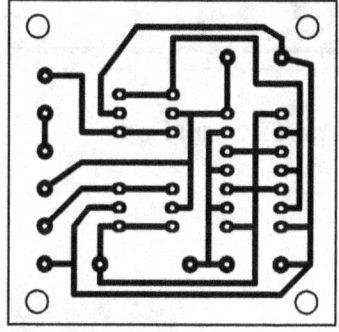

page 46 Dog Whistle

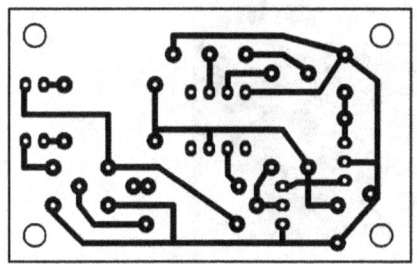

page 47 Coffee Thermometer

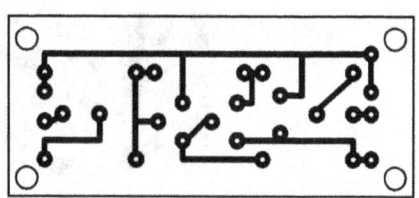

page 52 Cable TV Amplifier

Appendix

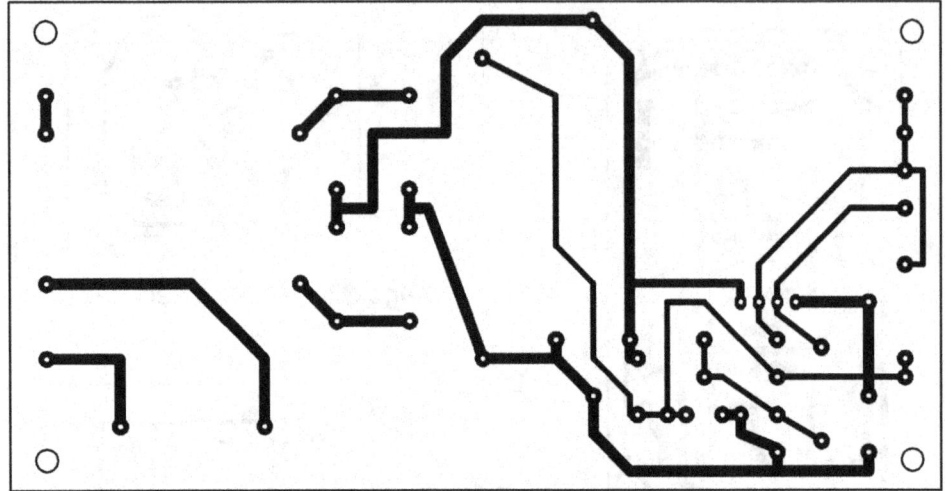

page 39 Drill Speed Regulator

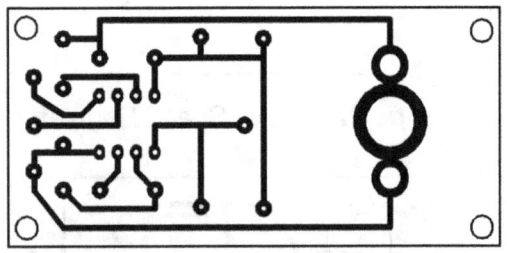

page 59 Mini Radio

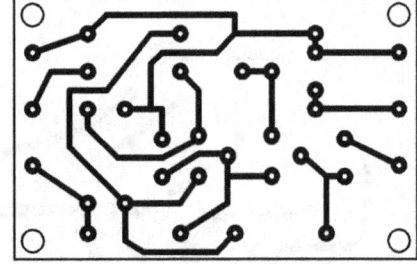

page 53 2-Transistor Radio

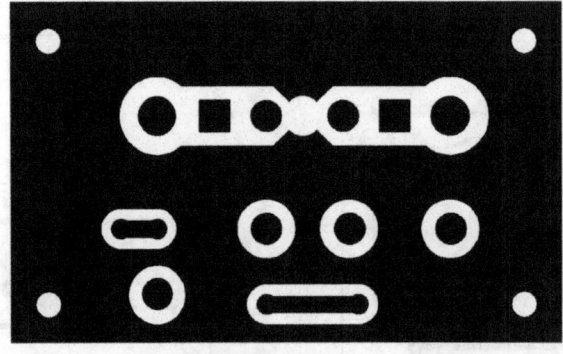

page 65 UHF Antenna Amplifier

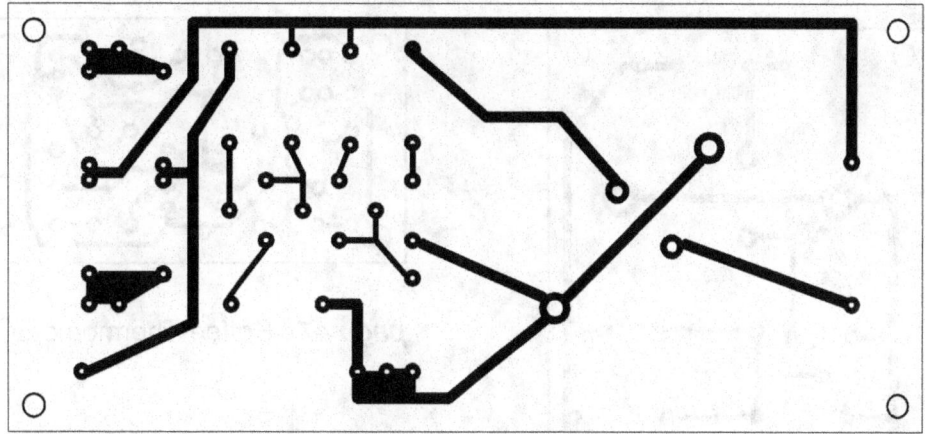

page 68 Nicad Battery charger

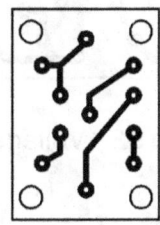

page 70 Signal Generator

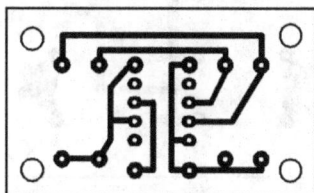

page 73 Superzener

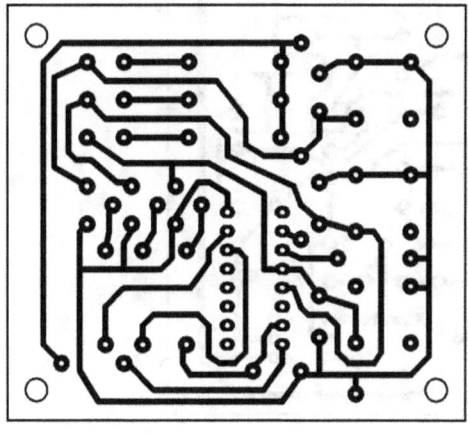

page 79 Selector Switch

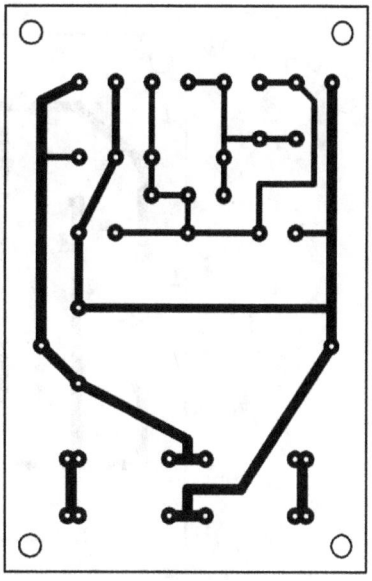

page 82 Power Supply

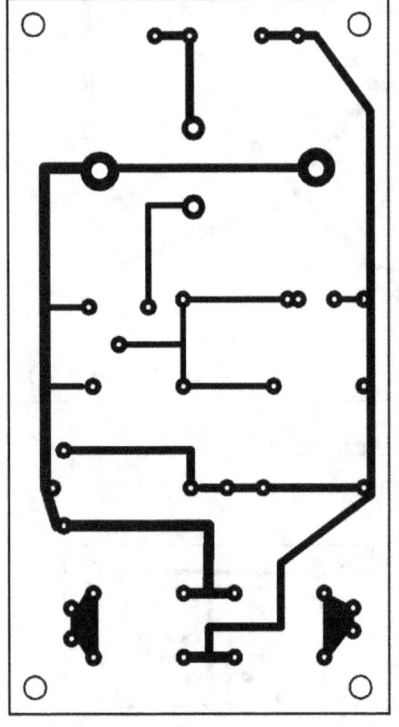

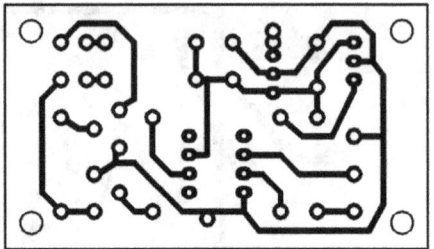

page 47 Coffee Thermometer #2

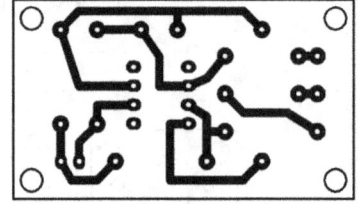

page 93 Voltage Monitor

page 83 3-Watt Power Supply

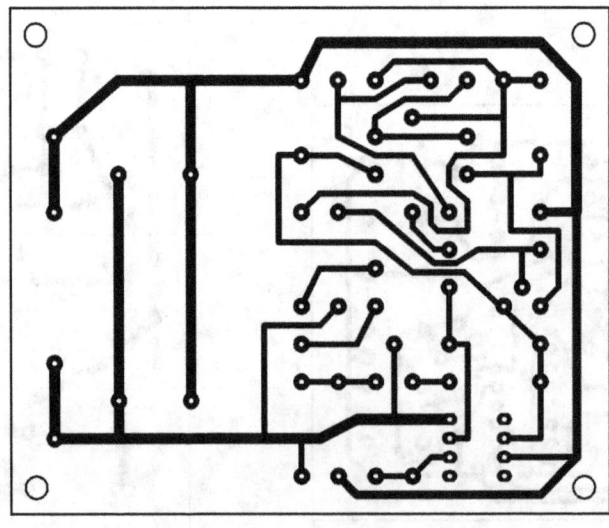

page 87 Nicad Charger

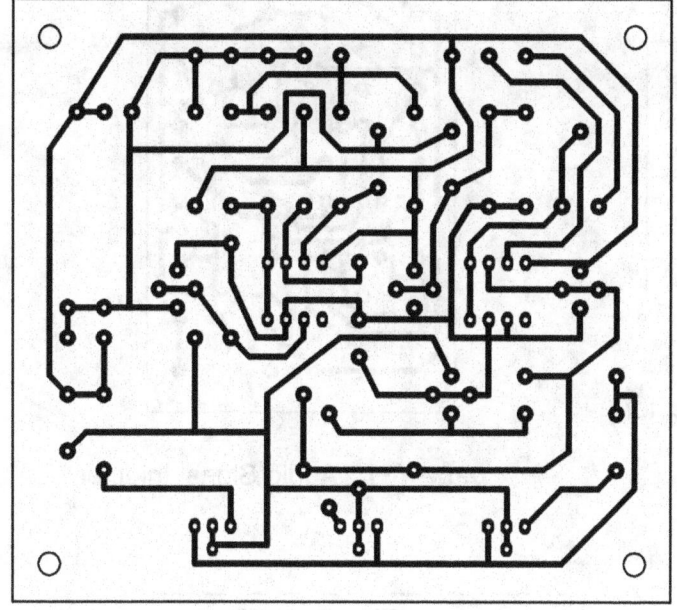

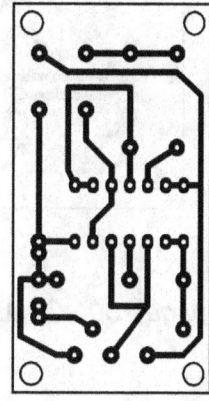

page 135   Servo Tester

page 90   0..50V/0..2A Power Supply

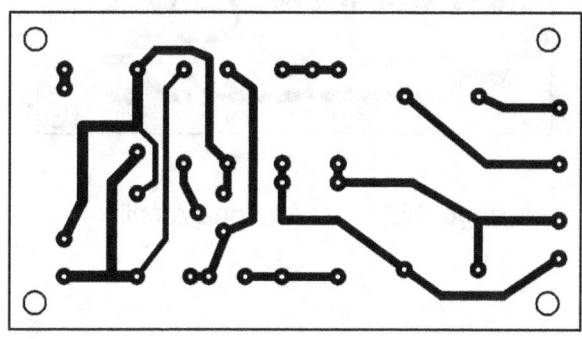

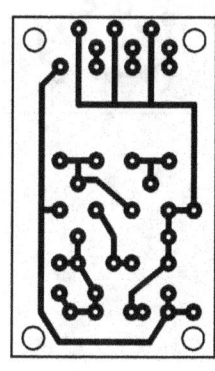

page 95   Current Alarm

page 112   Video Signal Mixer

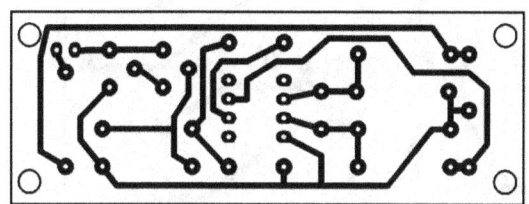

page 98   78XX Regulator Monitor

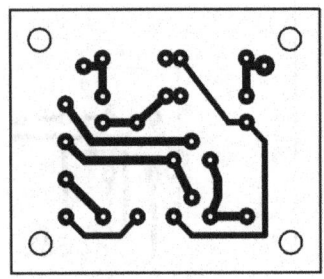

page 130 Continuity Tester

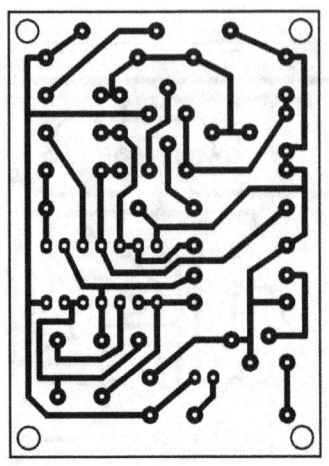

page 131 Audio Signal Injector

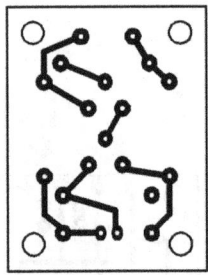

page 129 Lamp & Fuse tester

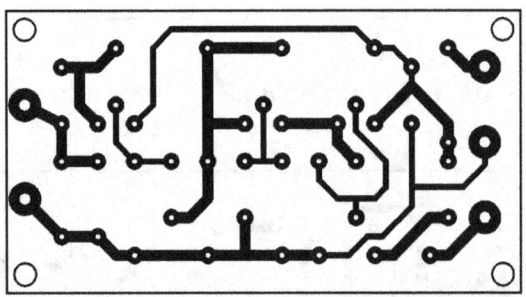

page 149 Car Antenna Amplifier

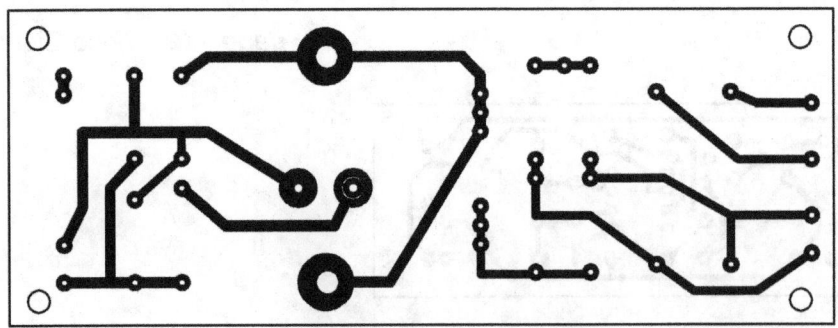

page 95 Current Alarm #2

This page is intentionally blank.

This page is intentionally blank.

Index

2N3055 26

A

A/D converter 109, 110, 137
alarm 45, 96
AM 53, 59
amateur radio 58
amplifier 26, 44, 52, 65
analog switch 31
analog voltage 110
antenna switch 56
anti-keybounce 79
AUDIO FILTER 23
audio mixer 15, 19
AUDIO WATTMETER 28

B

bandpass filter 23, 55
baudrate 102, 105
bridge oscillator 120
bucket brigade delay 74

C

CA3130 45, 85
CAR STEREO BOOSTER 12
carrier frequency 55
charger 68, 87
charging phase 88
Clapp oscillator 39
clipping 17
clock generator 118
clock signal 114
comparator 76, 100
converter 99, 105, 109, 124, 125
counter-EMF 37, 43
crossmodulation 62
cut-off frequency 23

D

darlington 84
detector 38, 138
DIMMER 36
division factor 71
DRILL 39
dynamic compression 17

E

echo effect 74
electronic fuse 94

F

FET 141
FET AUDIO MIXER 15
field strength meter 58
filter 55
FM 44, 62
FM modulation 134
frequency doubling 77
frequency meter 132, 141

G

generator 102, 119
GUITAR PREAMP 24

H

HEADPHONE AMPLIFIER 33
HF interference 44
HF-amplifier 53

I

intercom 44
IR sensor 138

J

joystick 104

K

keyboard 107

Index

L

LF356 25
LIGHT DIMMER 36
LM10 90
LM1014 93
LM2896 12, 13
LM35 50
LM382 44
LM384 44
LM3915 28
loadspeaker 21
logic probe 126
LOUDSPEAKER PEAK INDICATOR
     21
LOW-NOISE PREAMP 18
LTC1042 100

M

meter 58
MICROPHONE 32
microphone preamp 32, 72
mini-drill 39
mixer 112
monostable multivibrator 71
Morse code 63
MOSFET 58, 61, 64
MOTOR 37
motor 42
MOTOR SPEED 37
MOTOR SPEED REGULATOR 37
multimeter 138
multiplexer 78, 106
multivibrator 123

O

ocean emergency frequency 60
oscillator 46, 54, 61, 113, 114, 120

P

photo-coupler 49
piezo-ceramic element 46
piezo-electric crystal 45

PIN diode 56
potentiometer 78
power supply 82, 85, 90
PREAMP 32
preamp 25
preamplifier 18
preselector 64
PRESENCE DETECTOR 38
printer 108
printer interface 112
pulse generator 119

R

radio 53
radio-controlled 58
receiver 59, 61
reed switch 145
REGULATOR 37
regulator 37, 39, 48, 93, 98
RF amplifier 54
RF-amplifier 52
RS232 103

S

SAD512 74
sawtooth 122, 125
second-order filter 23
selector switch 79
sensor 139
servo 135
shortwave 55, 64
signal generator 70
signal injector 131
signal switchboard 29
sinewave signal 118
SPEECH PROCESSSOR 17
speed 42
SPEED REGULATOR 39
speed regulator 93
squarewave 113, 118, 122
SSM2015 32
SWITCHBOARD 29

Index

T

tachometer 145
telegraphy 63
telephone 49
temperature 45, 47, 50
temperature sensor 45
tester 130, 140
thermometer 47
thyristor 70
TIC 246M 42
timer 41
TL061 138
transistor tester 133
transmitter 134
triac 37
trigger 142
tuner 54

U

UA78G 82
UA79G 82
UHF 56, 65

V

VFETS 39
VHF 56, 64
video 112
voltage converter 124
voltage regulation 91
voltage regulator 48
voltmeter 128

W

WATTMETER 28
wireless 44

www.ingramcontent.com/pod-product-compliance
Lightning Source LLC
Chambersburg PA
CBHW081121170526
45165CB00008B/2516